How to Install
Automotive Mobile Electronic Systems

By Jason Syner

motorbooks

To Roger Syner for teaching me to be driven,
and to never stop pushing. —J.S.

First published in 2009 by Motorbooks, an imprint of MBI Publishing Company, 400 First Avenue North, Suite 300, Minneapolis, MN 55401 USA

Motorbooks titles are also available at discounts in bulk quantity for industrial or sales-promotional use. For details write to Special Sales Manager at MBI Publishing Company, 400 First Avenue North, Suite 300, Minneapolis, MN 55401 USA.

To find out more about our books, join us online at www.motorbooks.com.

ISBN-13: 978-0-7603-3177-4

Publisher: Zack Miller
Editors: James Manning Michels and Jeffrey Zuehlke
Creative Director: Michele Lanci-Altomare
Design Manager: Brad Springer
Layout by: Trevor Burks

Printed in China

On the cover: A well-planned system can look like a factory installation. *Brian Gomsak*

Inset: Aluminum or copper bar stock can be used to connect multiple capacitors.

On the title page: This is the interior of a Honda Civic Si that has cloth seats, carpeted floors, and a soft headliner with sound damping underneath. The painted door panels and other painted interior spaces add some reflections to keep the car from sounding too acoustically flat. *Joe Greeves*

On the back cover: The detailed wiring of an audio crossover and an EQ in the back of an SUV.

About the author:
Jason Syner has spent the last 15 years developing vehicles with intricate audio and video systems. He began in 1995 with an S10 truck loaded with Alpine equipment and has been working tirelessly in the industry ever since. Jason's 1995 Honda Civic, an artistic sculpture of flowing fiberglass and extreme paint, contained 11 Sony XPLOD amps, 10 Sony television monitors, and a rocking audio system. It won top expert awards in both IASCA and USACi, as well as awards for lighting and design from MECA. Most recently Jason partnered with Sony for another Honda, this time a 2005 Civic Si, which after its first trip to the SEMA show in Las Vegas earned Jason 14 magazine covers and top awards for sound quality, bodywork, and unique design. Jason is currently running a full-scale audio/video shop that provides car audio services as well as home theater design services. With his current car projects, Jason is developing new techniques for integrating Web 2.0 devices for mobile audio/video systems.

Contents

Acknowledgments

I want to first extend my thanks to the people who over the years have showed me the trade and placed some faith in me: Chris Yato, Ben Oh, Brian Schmitt, Gary Biggs, Steve and Maggie Head, Manville Smith, Gary Bell, Terry Floyd, Gord Stansell, Larry Chisner, and Ron Buffington. Thanks to Tony and Shelbi Pasquale for the kind introduction to this crazy industry, and Rob Liss, Michael Bryceland, and Rick Kojan at Sony for taking a chance on a little guy building cars at his house. To Chris and Melissa Owen and Chris Yato—thank you for the long hours on the phone and all of the countless words of wisdom.

To my family and friends—thank you for patiently putting up with my quirks and for letting me miss a birthday or two. Your support is appreciated more than you know.

To all of my sponsors over the years, including Sony, Meguiar's, Monster Cable, JL Audio, Mobile Solutions, Optima, Dynamat, Bob Myers, Joe Greeves, and all of the other 40 companies that have shared their wonderful products with me—many thanks!

My sincere gratitude to the people who have made this project possible: Adina Nichols, whose instructional design helped make this book a reality, and Jim Michels and all of the great people at Motorbooks; thanks for making this a painless process.

Also, to Angel Sandoval and Jose Brownes—thanks for the Xtreme Kustom Paint! Your tireless efforts made the Si a breathtaking masterpiece. Thank you. And a big thanks to Metodi Gushkov and the Sony Xplod team Bulgaria. Your emails and support have been an inspiration. To my most loyal fan (and friend) Mr. David Smith, thank you for traveling with me to so many car shows!

I want to send a very special thank you to all of the people who have volunteered countless hours to help me get my projects completed over the years: Roger Syner and Theresa Syner, Kevin Myers, Jeremy Briel, Chris Adams, Ryan, Tim Syner, Liz Syner, Katie Syner, Nick and Donna Nichols, Bobby Hilgaertner, Michael, and Steve Taylor.

Introduction

Whether you're upgrading a daily driver so that you can feel an intense boom in your trunk, or you're a competitive audiophile creating a sound quality (SQ) system—all audio/video installers need to have certain basic skills and knowledge. There are tools and techniques, tips and tricks, and basic rules of engineering and safety that must be learned and followed.

In the world of mobile electronics, you must have an understanding of electronics, woodworking, metal fabrication, fiberglass, sewing, drawing, and design and engineering. So what do you do if you're not a maverick supersonic that naturally has all of this knowledge? My advice is to read this book.

Car audio and electronics is a passion of mine. I started drawing my own concept cars at the age of 5 and have kept on sketching ever since. I built my first show car at the age of 15. I remember the smell of the fiberglass and the tingle of sawdust in my eyes—and I was hooked. My best memory is the first time I stayed up five straight days and nights to finish my first car show project. I was exhausted, hungry, and frazzled. But when I turned the key in the ignition and sound poured through my speakers, I was rewarded a hundred times over. It's the best feeling in the world to work hard at something, see it through to the end, and then reap the rewards of a job well done.

When I first started to learn about car audio and installation, I was like a leech to anyone who had more knowledge than me. I wanted to suck all of that information out of their head, and then turn around and figure out how to improve my talents utilizing some of their skills. What I didn't have was a guidebook or some sort of reference to help me through the challenges when no one else knew how to solve the problems. This is why I wanted to write this book—to give someone else out there the benefit of learning from my successes and the ability to avoid my failures. All of the content in this book comes from my own real-life experiences: problems that I solved, mistakes that I made, and victories that I won.

This book is designed for anyone at any level of expertise. If you're new to audio engineering, you can use this book to work on a car project from start to finish. But if you're a seasoned professional, use this book as a reference tool to help you work faster and more efficiently at your trade.

What's important to remember is that mobile electronics can be hard work. It's not a hobby you can pick up overnight, and it's not a craft you can be careless about. It takes careful planning, attention to detail, and commitment to see a project through to the end. Rest assured that the unexpected will always happen, things will inevitably take longer than expected, and you will have frustrating failures. But at the end of the day, when you've worked hard and overcome the challenges, you will be left with a vehicle that sounds amazing and work that you can be proud of doing.

So if you love the smell of fiberglass resin in your hair and enjoy listening to the purr of a 60-gallon air compressor—then I hope you enjoy reading this book just a little bit, too. I certainly enjoyed writing it.

Chapter 1
An Overview of Mobile Electronics and Basic 12-Volt DC Theory

If you made the decision to start a mobile electronic system in your car, then you've probably started thinking about what you want the system to consist of in the way of electronics. There's a lot out on the market to choose from—speakers, amplifiers, source units, accessories, and so on. Each type of electronic device has various components that interact with other types of electronic equipment in different ways. For example, component set speakers will interact with an amplifier differently than with a coaxial speaker. Despite the differences in electronic parts, they all follow the same set of rules.

Before you can start any kind of a mobile electronic project, you need to have a basic understanding of electronics, electrical theory, and electronic engineering. As we go through this chapter, we will discuss several common types of electronics and how you can apply basic **DC**, or direct current, theory to each of them. All of the wiring and electrical theory we will discuss later in the book will build on the information in this chapter. We'll also take a deeper dive into DC theory in Chapter 5.

ELECTRONIC ACCESSORIES (SATELLITE RADIOS, GPS, RADAR DETECTORS, ETC.)
Most of the electronic accessories that you'll want to install into your vehicle will be aftermarket accessories. These accessories are usually plug-and-play and are self-explanatory. Some of the more difficult ones that require setup will be covered in a later section of this book. Another type of accessory is **OEM** (original equipment manufacturer)—in other words, components that were specified by the

Most mobile electronic accessories get power from a cigarette lighter connection. Many modern cars no longer have cigarette lighters, but they still have the power adapter for your mobile accessories. Even better, some new cars have multiple adapters throughout the car for your many mobile accessories.

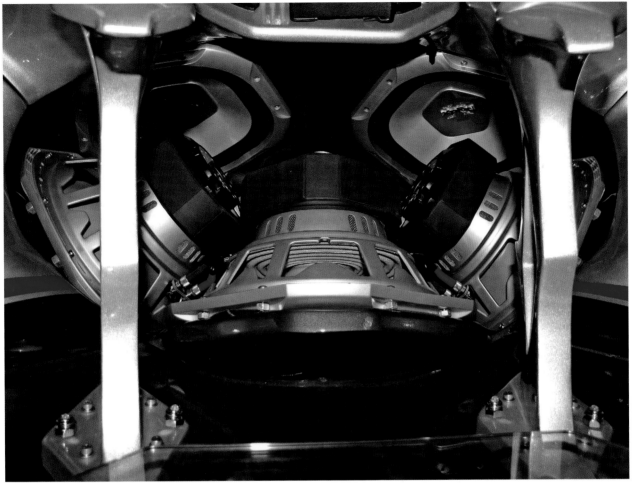

This free-air subwoofer enclosure holds six subwoofers. Two of the subwoofers mount in the traditional way with the cone exposed, whereas the other four subwoofers are inverted with the magnets showing and the cones playing into the box. When you do this, you need to invert the polarity on the inverted woofers. This means you wire the subwoofer to the amplifier with the negatives on the positives and the positives on the negatives. This box is sealed off between the passenger cabin of the car, and it's utilizing the trunk as the enclosure.

manufacturer of the vehicle (think GM, Toyota, Volkswagen, and so on) specifically for use in its vehicles. These can be purchased from a car dealership and are more expensive than aftermarket accessories. This is because they have a better resale value than an aftermarket product and are often of a slightly higher quality.

For example, you can pay as much as $2,500 for a Honda Accord OEM Alpine Navigation system with DVD. You could potentially buy a similar non-OEM product also made by Alpine for half as much, but chances are good it will not have the same features. It really doesn't matter which type you go with, and often you will probably find your budget helping you make your decisions.

Almost all aftermarket accessories require a standard cigarette lighter connection as the primary power source. **Power** is the rate of energy transmitted. It is the amount of current times voltage, measured in watts. **Wattage** is the amount of power, measured in watts. It is best to use a switched cigarette lighter connection that shuts off when the vehicle key is removed from the ignition. This keeps the vehicle's battery from being drained by the accessory while your car is off. There's nothing worse than killing your battery with an aftermarket accessory and having to get a jump start to get back home.

SUBWOOFERS AND ENCLOSURES

Adding a subwoofer as a complement to a stereo system will offer a giant improvement in the low-frequency energy inside a car with most of your musical selections. I'm sure that most of you have seen (or indeed, heard) a car pull up to a red light with a subwoofer booming. A subwoofer can add much enjoyment to a daily commute. There are several different types of enclosures that can house a subwoofer. These include a sealed enclosure, a ported enclosure, a band-pass enclosure, and a free-air enclosure. We'll cover these in more depth in Chapter 7.

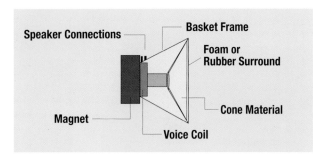

Here is a diagram of the various parts that make up a speaker/subwoofer. As you can see, its heart is a magnet, where the electromagnetic field is created. The magnet pushes the cone, which in turn moves the air. The voice coil receives the current and causes the cone to vibrate. The current comes into the speaker motor assembly through the speaker terminals.

wire or conductor, measured in amps) flows through the voice coil of the speaker. As current flows in one direction, it creates a magnetic field that alternates in the opposite direction. This field is attracted and repelled by the magnet, causing the speaker coil to vibrate, which in turn causes the speaker cone material to vibrate, creating sound waves. (This is the way all speakers we will discuss in this book function.)

Some subwoofers come with dual voice coils—which means that they have two separate voice coils, each with its own connections on the same cone. These subwoofers offer multiple wiring configurations to get more power out of your amplifier.

A subwoofer plays the low octaves of the music scale from 20 to 100 **Hz**. Hertz is a measure of sound frequency in time, or the number of cycles per second. A frequency of 1 Hz is equal to 1 cycle per second. It describes the musical presence of energy in this frequency range. Subwoofers come in the following sizes: 8, 10, 12, 15, and 18 inches.

Subwoofers are made out of a variety of materials, such as paper, Kevlar, polycarbonate, aluminum, and ceramic. The **baskets** (or frame and mounting flange) of subwoofers are constructed out of metals, such as aluminum, steel, and titanium, and usually have a gigantic magnet attached to the basket as well as an oversized voice coil. Alternating **current** (what flows on a

A new technology on the market today is the compact subwoofer, as seen here. It is tiny, and the magnet and basket are very thin so that you can mount the subwoofer in a small space. But if you're worried that this thing is too small to deliver sound—don't. The little subwoofer packs a powerful punch. The picture on the top is a Sony low-profile subwoofer that has a custom cone, designed to increase the sound wave output. The picture on the bottom is the same subwoofer with the high-power magnet built into the subwoofer's basket. These types of subwoofers can be purchased for around $100 and can be really useful if you are putting your woofers in a trunk but still want a lot of useable space.

On the left is a picture of a factory speaker. You can see the flimsiness of the cone. This is what causes them to blow so easily. On the right is a picture of a sturdier, aftermarket speaker, which tends to last longer and blow less often.

SPEAKERS

Speakers function in the same way as a subwoofer in that they transport sound through a cone and also in the way that they are constructed. However, nonsubwoofer speakers can play higher ranges, or **octaves**, of music. (An octave is the interval between musical pitches with half or double the frequency.)

Speakers come in a variety of impedances and can be wired to represent different **impedance** (or a speaker's resistance in ohms) at an amplifier. Speakers come in 2, 4, 6, or 8 **ohms**. (Ohms is a measurement of **resistance** or how much current will flow through a component.) Basically, ohms are how impedance is measured. Speakers' resistance will change while they are playing music. The speaker's ohm rating is usually given when the speaker is at rest (the cone is not moving). The higher the ohm rating, the less resistance and efficiency you will find in a speaker. A 2-ohm speaker will play louder but not sound as good, whereas an 8-ohm speaker will play cleaner and sound better. **Note: Ohm's Law** defines the relationship between power, voltage, current, and resistance. I suggest studying the relationship in greater detail (try the *Master Handbook of Acoustics* by F. Everest) to assist you in understanding speaker power and functionality as it relates to electronic engineering.

Most OEM car manufacturers spend the majority of their product planning concentrating on the vehicle's aesthetics, performance, and safety. They spend a slight amount of time

on the car's interior and comfort, but virtually no time on designing or implementing a high-quality sound system for your enjoyment. They try to spend as little time as possible in the research and development of the interior acoustics or sound quality inside a car. Because of this, the speakers that are provided to you with a new car purchase are usually poorly constructed and can only handle about 15 watts of power. (Please note, this is a generalization and is not specifically true of all car manufacturers.)

Speakers can function poorly inside a vehicle, so detailed planning is essential when choosing the proper speaker or speaker arrangement to use inside your car. Most speaker options provided by manufacturers are full-range drivers, which play all midrange and high-frequency information out of a paper cone. Some of the more expensive manufacturers in high-end models of certain cars provide a component speaker system for their cars. This system consists of a midrange speaker and separate tweeter speaker. The tweeter is typically mounted in the sail panel above the door panel or in the A-pillar. The midrange speaker is usually mounted at the bottom of the door panel. The tweeter will be aimed at the passenger's ears at a 45-degree angle to improve the high-frequency response. Because high frequencies are directional, pointing the tweeter toward the ear will make higher frequencies sound clearer. The midrange is usually mounted flat to

On the top is a picture of a coaxial speaker with the components all in one. In the middle is a picture of a component set, with all of the components as separate pieces. At the bottom is another set of component speakers shown mounted in the bottom of a custom door panel. Both types of speakers work well in different situations, and both make a car sound awesome.

the inner door skin. Some auto manufacturers provide a digital equalization of the sound inside the car. Most auto manufacturers have Digital Signal Processing (DSP) built into them. DSP uses complex algorithms to improve the sound within the car.

In the aftermarket world, it's possible to improve upon what the manufacturer has done with its sound system. Various types of aftermarket speakers are available. For example, a coaxial speaker is a single midrange speaker with a tweeter mounted in the center of the cone. This driver is the most popular choice for aftermarket shoppers because you get the two separate drivers together for the price of one speaker. This speaker is very easy to install, as long as you've purchased the same size coaxial speaker as your original factory speaker. There is a capacitor wired to the tweeter, which only allows tweeter frequencies to play out of the speaker. Also, most coaxial aftermarket speakers are better built and have a bigger magnet than a factory speaker. This enables them to dissipate more heat and play louder.

Another choice for aftermarket speakers would be a component speaker system. This system consists of a midrange, tweeter, and **passive crossover network**. A passive crossover network is an electrical circuit made up of a coil, capacitor, and resistor. It takes the signal from the head unit and breaks it into sections, sending the high frequencies to the tweeters, the midfrequencies to the midrange speakers, and the low frequencies to the midwoofers. It handles a lot more power than a factory speaker system, anywhere from 75 to 200 watts, compared to 10 to 50 watts in a factory system. Even if your vehicle comes with a component set of speakers, chances are good that they are not going to have the sound quality or the power capability of an aftermarket component set.

SOURCE UNITS

One of the most important parts of your stereo system is the source. In this book we'll refer to the electronics that generate the music as source units. Aftermarket source units include CD players, and most have iPod control functions. Source units provided by manufacturers are typically CD players and sometimes CD players with navigation screens attached to them.

Aftermarket source units offer a lot of options that your factory stereo system does not, such as iPod control, equalizers, CD text, and the ability to read MP3 data disks. They also have a higher quality playback for the source material. Most aftermarket source units have a digital analog converter (DAC). To explain, a CD player plays a disk of digital information in 1s and 0s. A DAC converts the 1s and 0s that the ear cannot hear to an analog signal voltage that is output through the RCA jacks on the back of the source unit, or through the speaker output wires. The higher the quality of the DAC, the less digital information is lost or colored, and thus the better your system will sound. The

Top: a single DIN. Bottom left: a DIN and a half. Bottom right: a double DIN. Make sure you always measure properly for the DIN space you have in your vehicle before shopping for your source unit.

more 1s and 0s that are not lost or colored, the more analog information is output by signal voltage. To put it simply, with a higher quality DAC there is more musical detail in a recording available for listening.

Source units come in the following sizes: single DIN, DIN and a half, and double DIN. A DIN is a German measurement for in-dash players that will install in a single slot. The dimensions for a rectangular DIN opening are 7

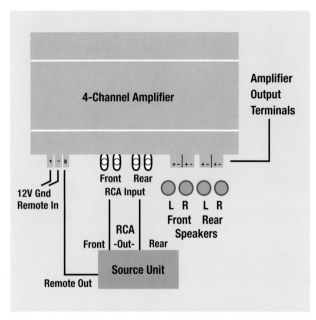

This is a schematic of a four-channel setup with front and rear speakers. For this type of design, you would connect a power wire to a battery under the hood, which would connect to a fuse, also under the hood. The fuse protects the amplifier from a short on the power wire. The power wire then runs out from the fuse through an opening in the firewall to the 12-volt terminal on the amplifier. The same size cable connects to the negative terminal on the amplifier and runs to a seatbelt bolt for a ground. On the back of the source unit, a blue remote wire connects to the amplifier's remote terminal.

inches wide and 2 inches high. Thus the measurement for a DIN and a half is 7 inches wide and 3 inches high, and a double DIN is 7 inches wide and 4 inches high.

Source units offer specifications that should be considered when making your purchase:

- **Signal-to-noise ratio:** Compares music level to background noise level; the higher this ratio the better. The signal-to-noise ratio of a high-quality source unit would be 109. To explain this further, on this particular CD player, the noise present in the background would be one decibel, or 0 **dB** and the signal would be measured at 109 dB. (A decibel is a measurement of sound volume.)
- **Channel separation:** Usually measured in decibels at one kilohertz (**kHz**). Any CD player between 95 and 100 dB is considered exceptional. Anything above 80 dB would be acceptable.
- **Frequency response:** The range of frequencies that the source unit is capable of playing. A safe, general requirement is 20 Hz to 20 kHz. This is the basic audible range of human hearing.
- **Voltage preout:** The amount of **voltage** (what pushes current through a circuit) output on the RCA jacks. The higher the voltage, the better the sound quality and dynamic range of the system. The average would be 4 volts, with 8 volts above average.

AMPLIFIERS (AMPS)

No stereo system would be complete without a high-power amplifier. Amplifiers work the way their name implies—by amplifying the signal coming into them and outputting it to your speakers. It does this by taking current from the power supply and sending out a signal to match the input signal but with larger amplitude (or waveform). It is very important that your amp be linear—it should input and output the signal uniformly. To keep the signal uniform, you don't want to run a small amount of input signal to an amplifier, and you can compensate for this by turning up the output signal gain adjustment. You need to plan on running the same amount of signal into the amplifier as you plan on running out of it. **Note: Amp**—can be used two ways: short for amplifier, it is an electronic component; short for ampere, it is a measurement of current.

There are many different brands of amplifiers, but there are really only three types: Class A, Class AB, and Class D. Most amps are transistor amplifiers. They have transistors in their output stage to increase their power output. Class A amps' output transistors are always conducting current. This makes them inefficient because they use a lot of DC voltage and get extremely hot. In a Class A amp, 100 percent of the input signal is used. However, these amps are known to have a very warm sound. Class AB amps use two output transistors; each transistor only amplifies half of the input, making them more efficient. These amps have a more neutral sound. Class D amps are the most efficient and are often referred to as the switching amps. These amps use pulse-frequency modulation, which means the input signal is converted to pulses. The pulses are increased to match the magnitude of the input signal and output to the speaker. Very little power is lost by the transistor, due to the on-and-off state of the pulses. Class D amps sound best on subwoofers, but they also have a lower damping factor, meaning it is affected more by a loud speaker than an amp with a higher damping factor. (See specs below.)

Amplifiers also come in a variety of configurations: multichannel, stereo, and mono. Multichannel amps are usually four to six channels and can be set up in many different ways. For four-channel amps this would include putting two channels each on the front and rear (which amplifies your front and rear speakers); putting two channels on the front speakers and two on the trunk-mounted subwoofer; or in high-end sound quality systems, putting two channels on midrange speakers mounted in the kick panels and two channels on midbass drivers mounted in the front doors. Six-channel amplifiers can be done the same way but would have the remaining two channels bridged (or combined) for the trunk-mounted subwoofer. This doubles the power coming out of the amplifier. Mono amplifiers play one mono-output channel and would only be utilized in one of two configurations: (1) mono onto a single or pair of subwoofers (if it had enough power to handle two speakers), or (2) mono onto a single driver in the front left speaker and another mono on a driver for the front right speakers (creating stereo sound with two amplifiers, giving you the best channel separation that money can buy). Finally, stereo amps are simply two-channel amps designed to run two subwoofers in stereo, two midrange speakers, or two midbase speakers.

When choosing your amplifiers, buy from a reliable manufacturer such as JL Audio, Sony (economical yet reliable), JBL, or Alpine. (Gary Biggs has won more world championships for sound quality using JBL amplifiers in his Buick than any other brand or competitor.) The reason for doing this is that quality amplifiers will always improve the sound in your vehicle. If you buy a no-name or low-end brand, then you are seriously sacrificing the quality of your sound system. Amplifiers are never items that you want to pinch pennies on. Another thing you must consider before buying your amps is your plan for selecting the proper number of channels for your system. I cannot stress to you enough the importance of planning, planning, and planning for your system long before you buy or work on anything. You must select the number of channels you'll need, or you'll end up exchanging your amplifier before you ever finish your system. Finally, make sure you select an amp with a solid manufacturer's warranty so that your investment is protected.

On the top-left is a factory battery from a car that has 550 cranking amps, which is relatively weak. This battery can only be drained three or four times before it will no longer hold a charge. The battery on the top-right is a starter battery that has 1,000 cranking amps, which is almost double a factory battery. The battery below on the left is a deep-cycle battery and has the longest run-time of any battery. This battery can usually run a couple of hours before going dead and can be drained and recharged repeatedly without damaging the battery. On the bottom-right is a capacitor, which is similar to a battery but rather than discharging slowly, it can discharge almost instantly, making it an excellent component for providing power to an amplifier or any other high-current electronic component.

This Sony in-dash digital equalizer allows you to cut or boost frequencies to improve the sound quality inside the car. The frequency is selectable and so is the level in the Q. You have a total of 10 bands that you can adjust.

"Plan, Plan, Plan"

With amplifiers you also need to consider specifications. These include but are not limited to:

- **Damping factor:** Tells you how the amplifier controls the speaker. An average damping value is between 250 and 500. A more high-end damping value is between 800 and 1,000. An amplifier with a lower damping value will be affected more by the load of a loud speaker and can cause the amplifier output to change. The higher the damping value, the less affected the amplifier will be by the speaker load.
- **Frequency response:** Should be 20 Hz–20 KHz
- **Total Harmonic Distortion (THD):** Should be less than 0.05 percent at full power rating—the lower the THD, the better your amplifiers sound.
- **Signal-to-noise ratio:** Anything rated near 100 db is acceptable.

Wiring subwoofers to your amplifier can increase the amplifier's output, if you're using dual voice-coil subwoofers. A dual voice-coil subwoofer with 4 ohm impedance on each coil and wired with the coils parallel (meaning positive to positive and negative to negative) will give you 2 ohm impedance at the amplifier. This will give you a 50 percent power increase out of the same amplifier over using a single voice-coil, 4-ohm subwoofer. To illustrate, an amplifier that does 200 watts at 4 ohms would give you an output of 200 watts with a single, 4-ohm voice coil connected to it. That same amplifier connected to a dual, 4-ohm voice coil would give you an output of 400 watts at 2 ohm impedance.

Batteries and Capacitors

A car battery is a rechargeable battery that supplies energy to your car and to your car stereo system. A battery uses chemical reactions to produce electrons and allows them to flow through the conductors on the top of the battery to produce electricity. There are different types of batteries—mainly starting batteries (which is generally what you'll have in your car) and deep-cycle batteries. Starting batteries are designed to start your car. These batteries have higher cranking amps and produce large amounts of starting current for short periods of time. Deep-cycle batteries are designed to be charged and discharged. Optima offers some of the best deep-cycle batteries, allowing you to discharge them over and over again and still allow for additional recharging. You can discharge them down by 80 percent and they can still be recharged. On the other hand, when a starting battery goes down even 30 percent, it will not allow you to recharge it.

The best setup for a car with a sound system would be to have a starting battery under your hood to start your car and a deep-cycle battery located somewhere else to run your audio system. The deep-cycle batteries are best mounted somewhere in the trunk. Using a deep-cycle battery as a starting battery is possible, but it will offer fewer cranking amps. If you do this, you should select an oversized deep-cycle battery to compensate for the lower amount of current available for starting. I recommend the Optima Yellow Top D34 battery, offering 750 cold cranking amps and 120 minutes of reserved capacity.

A capacitor is an electronic part with a pair of conductors (positive and negative) separated by a dielectric insulator. Similar to a battery, a capacitor charges and discharges, but it does so very rapidly. What this means is that a capacitor has stored energy that can be discharged rapidly, which helps your car audio electronics' energy needs. Audio capacitors are available in ratings of 0.5 farad, or 1 farad up to 500 farad. A **farad** explains the measurement of the amount of storage inside a capacitor. The higher the farad rating, the more storage you will get out of your capacitor. (For more information, please see Chapter 5.)

Most musical selections have peaks in the recorded track. These peaks are present in the signal **waveform** that goes into an amplifier and will produce a strain on the amplifier's output. (**Waveform** describes the shape of a signal moving through a space.) Mounting a capacitor in close proximity to your amplifier (within inches) will allow a rapid discharge of current into the amplifier's power supply and will allow it to produce an output consistent with the musical piece. If needed, a capacitor can dump its entire charge in a split second.

Using a deep-cycle battery along with a capacitor gives you the best of both worlds for your system. The deep-cycle battery will allow you to run the system longer and run the battery down, and you'll still be able to recharge it. Meanwhile, your capacitor will help the system play louder and regain its composure faster.

Crossovers and Equalizers

A lot of source units' amplifiers include a crossover and an equalizer (EQ). You can also purchase an EQ and crossover as separate components to wire into your system. These components are some of the most important parts you can add to your system.

A car audio crossover is designed to split the audio signal into separate frequency bands to go to your loudspeakers. For example, a tweeter is a high-frequency driver, and it performs best by receiving only high-frequency signals. The crossover would then only send the high-frequency signals to the tweeter. Likewise, the midrange speakers would then get only the midrange signals, and the subwoofers would receive only the low-frequency signals.

When selecting a crossover, look for one with a steeper slope. Slopes are measured in multiples of 6 dB, from 6 dB up to 72 dB, with 72 dB being the steepest. The slope represents how sharply the crossover will turn on or off the particular signal for a specified speaker. Simply put, a crossover with the steepest slope has the best ability to separate and emphasize the signals within each of the appropriate speakers.

An EQ works by cutting or boosting particular frequencies. There are a lot of problems for sound reproduction inside automobiles, caused by all of the interior reflections in the small space. For example, shiny glass windows, a hard plastic dash and door panels, and sometimes leather seats can all be reflective surfaces. These reflections wreak havoc on your audio system because the sound bounces off of the hard surfaces and causes distortion. A lot of times reflections will cancel or distort the new sounds coming from your speakers. An EQ can cut unwanted frequencies or boost desired frequencies, helping to control the sound problems inside your car.

EQs come in two configurations: One-third octave and parametric. One-third octave EQs have 31 adjustable bands, plus or minus 6–9 dB. Parametrics offer 10 bands (usually), which are adjustable plus or minus 12 dB. In addition to being able to cut or boost a frequency, they enable you to adjust the Q sharper or wider (flatter) to pull an entire frequency region up or down, sharp or flat. Once you adjust a frequency band up (boost) or down (cut), you can also adjust the Q, which determines whether the adjustment is sharp (it affects only that frequency) or wide (it affects around it).

EQs and crossovers are very important in a quality sound system; they can fine-tune your sound quality and help you get the most performance from your system. However, these components also are the hardest to install and take a lot of skill to tune accurately. It will take you a lot of practice to learn how to listen and then make the proper adjustments to these components in your system. I recommend learning how to "tune" with these components with a seasoned professional, until you learn to hear the nuances on your own.

Video Monitors

Over the last decade, video has become increasingly popular inside cars. Road trips can be a lot of fun with the ability to watch the latest movie inside your car. If you have kids, you may have already discovered that it is impossible *not* to have video in your car.

Note: When putting video in your car, please remember that there are laws about where monitors can be placed. It is illegal to drive with monitors in the front seat of your car while they are turned on, so be sure to follow the law and place your monitors in places that are not distracting to the driver. In most states, video monitors in the front seat can only play video when the car is parked and the emergency brake is on.

Video monitors are available in three basic types: (1) flat-panel monitors for headrests, (2) flip-down monitors for overhead areas (these are larger in size and come with headphones), and (3) flip-out touch screens for in-dash source units.

Flat-panel monitors in overhead consoles or headrests require a DVD player to be installed inside the vehicle.

Source units with in-dash flip-out monitors have the DVD player already built into them. You can also use a video iPod to output movies to a monitor. Once you have video in your car, it can be a lot easier to integrate other types of electronic accessories, such as video game systems and navigational systems.

When picking monitors, make sure you select one with a good off-axis picture quality, because this will make it easier to view in different levels of lighting. Keep in mind that monitors are harder to view in daylight, so you might want to consider tinting your car windows.

The types of electronics that you can integrate into your personal system are only limited by your imagination and your knowledge of how to use them. I have seen car installers integrate karaoke machines, disco lights, digital musical instruments, and a plethora of other innovative and wild items. I encourage you to learn as much as you can about how different products work independently and with other products so that you can begin creating the next crazy, cool idea.

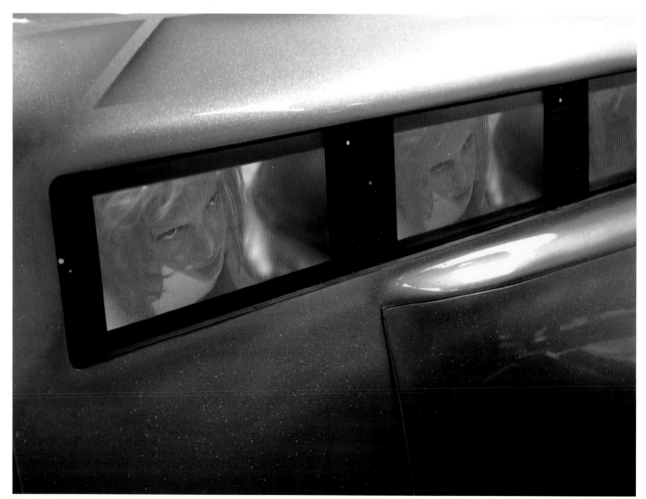

Here is a sample of what a finished set of video monitors can look like in a car. In this case, a group of 7-inch wide-screen LCD monitors were installed in a horizontal row in the back seat of a car. The six parallel screens created a 42-inch video display that could run DVD movies or MPEG graphics. *Joe Greeves*

Chapter 2
Mobile Electronic Accessories

If you're anything like me, then Christmas and birthdays are all about getting the latest gadgets and toys that are coming out. If it's new and hot, then I want it and I want to know how I can use it my car. This includes cell phones, game systems, navigation, computers, and iPods. I practically live in my car, so it's really important for me to be entertained and well-connected while I am out driving around.

As I mentioned in the first chapter, mobile electronic accessories can easily be installed with a cigarette lighter connection for quick plug-and-play. You can find affordable accessories to help you zip through traffic a little faster, or you can invest in high-end ones if you need your "James Bond" ego stroked a little bit. Let's take a look at some of the more common accessories you may want to use to complement your rocking sound system.

GPS

Quickly becoming a staple item in the average American driver's car is the GPS, or global positioning system. The more reliable, user-friendly, and consistent brands of GPS are TomTom and Garmin. GPS devices are supposed to help you find your way through traffic and give you directions to your destination—but remember, these devices rely on software updates. This means that periodically you must take them out of your car and hook them up to your computer so that they can keep current with changes to destinations, roads, and traffic patterns. If you don't update the software frequently, then chances are good it will eventually steer you the wrong way.

Some of the more advanced GPS devices give the ability to use a Bluetooth connection between your cellular phone and the GPS unit. When this Bluetooth connection is established,

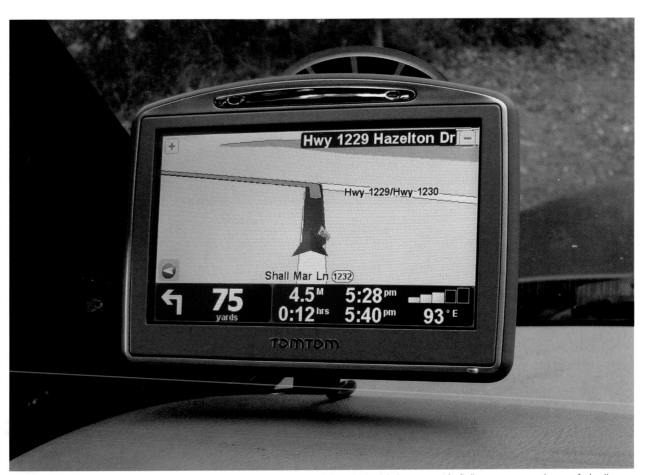

Here is an example of a common GPS unit, a TomTom, mounted to a windshield. These little navigational devices are great for finding your way around or even for locating specific businesses, such as gas stations and grocery stores.

you can accept and place phone calls through the GPS device and have the audio broadcast through the GPS device's internal speakers. Some GPS devices have an audio out connection. This connection can be connected to a car's auxiliary in connection, which is located on the dash of many new vehicles. Most GPS devices, including TomTom, will only let you transmit the cell phone call through the GPS' internal speaker. It will, however, allow you to transmit MP3 music and audio books stored on the GPS device through the car's audio system. Another way you can connect the GPS to the audio system is if your vehicle is newer and has an auxiliary (AUX) input on the dash. With this input, you'll need a mini, 1/8th-inch input plug. You'll then run the audio output from the GPS to this input on your dash. This is a relatively inexpensive cable, and you can purchase it at Radio Shack or Best Buy.

If this is a feature that appeals to you, then I recommend hard-wiring your GPS device to your car in order to help the audio streaming through the device operate more effectively. Most GPS units have a built-in modulator, or transmitter. This handy feature will broadcast your Bluetooth music (MP3s) through your audio system. You can do this by turning your FM dial to one of the lower stations (between 85 and 90 on FM) to pick up the audio stream on a dedicated frequency.

BLUETOOTH

You don't have to have a GPS device to install a Bluetooth connection in your car. There are other accessories available to help you talk and listen on your cell phone through your car's audio system. Bluetooth devices usually come as a plug-and-play kit that attaches to either your dash or sun visor. Motorola makes a decent kit, which can be bought for under $120. I do recommend making sure that your cellular device will work with your Bluetooth device before purchasing a kit. Many cellular stores can assist you with choosing the right in-car Bluetooth kit to work with your phone.

Your other Bluetooth option is the professional-grade hands-free kit called Parrot. Parrot makes several different models that range in price from $100 to $500. These kits enable you to play MP3 files, as well as send and receive cell phone calls through your audio system in your car while driving. Another cool feature of the Parrot kits is that they have a voice recognition system that you can use to dial contacts. When you accept any Bluetooth call through a Parrot or GPS system, you will have a caller ID function as well. This can help you decide if you even want to take the call while you're driving.

The Parrot kits are designed to be installed professionally and are hard-wired into a car. They are not designed to be plug-and-play installation. You could compare the wiring of these to that of an alarm system. It can be a lot of work, so if you're not comfortable doing this kind of work, let a retailer do it for you.

It's pretty safe to say that these days most new high-end vehicles come with an OEM Bluetooth integration system comparable or better than a Parrot system. If you're buying a Lexus, BMW, Range Rover, Mercedes, or Acura, you can count on having an ultra-elite, ultra-safe Bluetooth system in your car.

Let's take a look at how to set up a Bluetooth connection between a cell phone and a GPS unit.

This is a TomTom Go 920 GPS unit and a T-Mobile Blackberry Curve. You can connect these two devices via Bluetooth by following a few simple steps. Even if you don't own these particular models of GPS and mobile phones, the Bluetooth connection setup process will be roughly the same.

Step 1: On the Blackberry phone, start the connection process by going into the phone options menu and selecting Bluetooth settings.

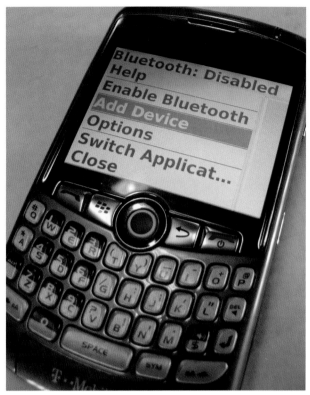

Step 2: Under the Bluetooth options, select "Add Device."

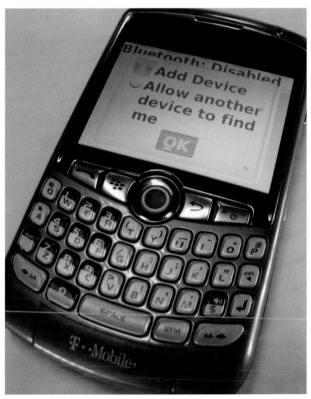

Step 3: The Blackberry will ask if you will allow another device to find it, and you should select "OK."

Step 4: The Blackberry will begin to search for available Bluetooth-enabled devices.

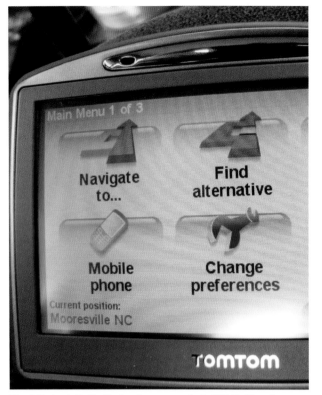

Step 5: Next go to the TomTom's options menu and select "Mobile Phone."

Step 6: The GPS device will ask you if you want to establish a Bluetooth connection between it and your phone. Press "Yes."

Step 7: The GPS will then ask if you want it to start searching for a phone, and you will also select "Yes."

Step 8: The GPS will search for a mobile phone once you have given confirmation to do so.

Step 9: The GPS and the Blackberry will begin to search for Bluetooth devices in unison. If this doesn't happen, turn off the devices, reboot, and begin again. You have to work quickly if you want the devices to search simultaneously.

Step 10: The GPS will ask for confirmation to connect to the Blackberry smart phone. Select "Yes."

Step 11: The GPS will begin a connection process to the Blackberry, which should take a couple of minutes to complete.

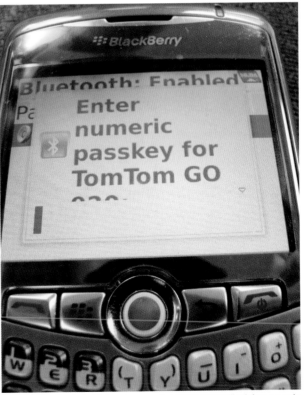

Step 12: Often the device that you're connecting by Bluetooth will ask for a numeric passkey. The default numeric passkey for most devices is 0000.

Step 13: The Blackberry will ask you if you want to accept the connection from the TomTom. Click "Yes."

Step 14: Once the devices are connected, you will see a confirmation screen showing that the devices are synched and ready for hands-free calling. In addition to being able to do hands-free calling and having access to the phone's address book by the GPS' onscreen menu, you will also be able to download weather alerts and real-time traffic information (provided your cell phone is equipped with a data plan.)

Step 15: Next the TomTom will ask you if you would like to copy your phonebook to the TomTom Go for easier hands-free dialing.

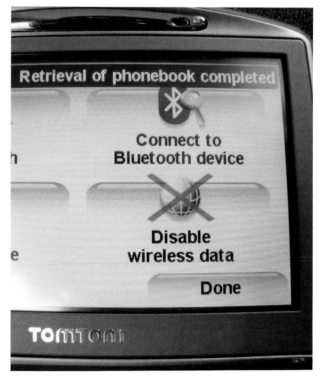

Step 16: Do not move either device until you have received the confirmation message that the retrieval of the phonebook is complete.

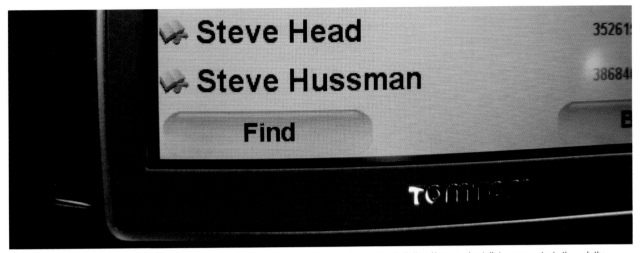

Step 17: The full Blackberry phonebook is viewable on the TomTom, and you will then have one-touch dialing. You can also talk to your contacts through the TomTom's internal speaker phone.

IPODS AND MP3 PLAYERS

Another popular device that has become a staple to most car audio systems is the iPod. These little MP3 players often hold your entire music collection—at least they do mine. I always hate when I get into a car that doesn't have an iPod connection, because that means you are at the mercy of looking for a CD or turning the radio dial. In this high-tech world, those are just not ideal options.

Many source units now come with iPod connection capability, although you do have to pay a little more for that feature. Likewise, you can buy a cigarette lighter adapter kit for any iPod (or any other MP3 device for that matter). The reason I like to connect my iPod directly to the source unit is that it eliminates a lot of the wires running across my dash. You can keep your driving space tidy while still using your iPod library. Pioneer makes an affordable yet quality source unit with an iPod connection (the DEH-P510UB), and it's really easy to use.

ALARM SYSTEMS

Before you go and add too many accessories to your car, I highly recommend adding an alarm system. Unfortunately, in the world we live in, you can never be too careful. If you are going to invest in a quality sound system and cool accessories, then you need to protect your investment. A decent alarm system can be bought for around $100 plus the cost of installation. Some retailers will install your alarm for free with the purchase of the alarm, or vice versa. The great thing about an alarm system is that it can give you power locks and remote keyless entry—even if you don't already have those features in your car. Clifford is the best alarm device money can buy, but Viper is a great choice because it provides supertight security at an affordable price.

You can purchase source units that allow you to connect your iPod directly to them. The owner connected this iPod to an aftermarket source unit. In many vehicles, this same connection may be located somewhere else on the dash; it is not always going to be found at the source unit.

Radar detectors are usually mounted on top of the dash, near the steering wheel or, as shown here, they can be molded into a custom fiberglass gauge-pod. *Joe Greeves*

Radar Detectors

Much like car alarms, radar detectors have been around for a while, and the technology just keeps getting better. Beltronics makes an awesome radar detector that delivers long-range protection on all radar bands, including X, K, KA, KU, and POP. Just keep in mind that a radar detector is not your free card out of a ticket. It is not an excuse to speed or drive like a maniac, and the police can still catch you. However, if you like the added security of knowing who is around you, then I recommend the Beltronics RX75 Remote. Just be sure to check that detectors are legal where you drive.

ACCESSORY ADAPTERS, POWER SUPPLIES, AND WIRING

You may be thinking that all of these accessories are going to clutter up your car. Well, you're correct. There are few ways to deal with the mess of wires. Most cars only have one or two power adapter plugs, and if there is more than one, they are usually not located next to each other. This means that if you need to utilize both of them you will have wires running to two different places. Instead of having your wires running all over the place, you can buy a multiplug adapter for your cigarette lighter connection and then zip tie the wires up in a nice, neat line.

The multiplug adapter for a cigarette lighter power supply converts one power supply into two adapters, so you can use more than one accessory at a time. These are pretty inexpensive and can be found at Radio Shack.

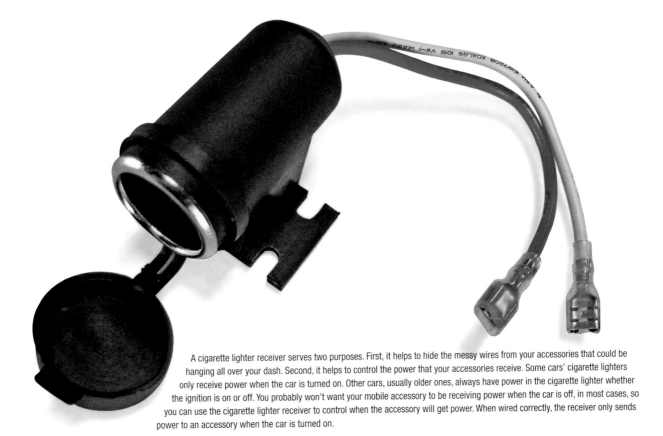

A cigarette lighter receiver serves two purposes. First, it helps to hide the messy wires from your accessories that could be hanging all over your dash. Second, it helps to control the power that your accessories receive. Some cars' cigarette lighters only receive power when the car is turned on. Other cars, usually older ones, always have power in the cigarette lighter whether the ignition is on or off. You probably won't want your mobile accessory to be receiving power when the car is off, in most cases, so you can use the cigarette lighter receiver to control when the accessory will get power. When wired correctly, the receiver only sends power to an accessory when the car is turned on.

These adapters can convert one power adapter to two or three adapters, depending on which one you buy. You can find these adapters at any electronics retailer. However, if you really want to make the dash look sleek and clear of the bird's nest, then I suggest hardwiring the mobile accessories directly to the car.

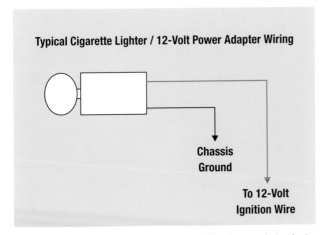

Typical Cigarette Lighter / 12-Volt Power Adapter Wiring

Chassis Ground

To 12-Volt Ignition Wire

This diagram (inset) illustrates how to hardwire an additional power adapter plug to your vehicle. Doing this will enable you to plug in devices that will turn on when you turn your key on and turn off when you turn off the key. The red wire is the power wire for the adapter and should be wired to the car's ignition wire that has 12 volts present on it with the key on and no voltage present with the key off.

All accessories plug into the cigarette lighter adapter, which provides 12 volts of power to the device. The plug that goes into the cigarette lighter connection has two prongs on the end. These prongs are ground connections. The center pin on the end of the plug is your 12-volt power. Most auto parts stores have a cigarette lighter receiver connection that is a female plug and you can plug your cigarette lighter cord into. This plug has positive and negative wiring that comes off of it. You'll want to take the negative connection and connect it to a chassis ground point (bolt) and tighten it down.

Note: The color and location of the 12-volt ignition wire can be found by consulting the service manual for your vehicle. Service manuals are available at dealers and most auto parts stores.

Then connect the 12-volt wire to one of the switched ignition wires coming off the steering column that will have 12 volts of power when the key is on. (When the key is off, the wire will have zero volts of power.) While at the auto parts store, be sure to purchase a low-amp fuse (around 5 amps) so that you can wire it in-line on the positive lead. Once the two wires are connected, tuck them behind the dash and feed up the power plug for your device and plug it in. Once you've completed these steps, you'll be able to turn on your car and have your GPS device automatically start up and be ready to go.

By wiring your mobile accessory directly to your car, you can hide most of the wiring. This leaves your dash is free of clutter, as with this radar detector and gauge cluster, with no wires showing. *Joe Greeves*

Here are four different types of LED accent lighting made by StreetGlow that you can purchase for your car. Left to right, the first is an LED light bolt, which is a bolt with a nut that you can use to secure an amp or speaker to your vehicle. The center of the bolt lights up with the color LED of your choice. The second item is an LED projector light that has an adjustable base that allows you to point the glow of the lighting to shine wherever you would like it to. The third items are LED singles, which are four individually housed LED lights that are tiny enough to place in very small places and still get a very bright accent glow. The last item is an LED scanner, which is small plastic housing filled with LED lights. These lights have nine different patterns and pulses, which flash to create different effects within your car.

ACCENT LIGHTING

Accent lighting is a great way to show off the little details in your system and to bring a unified style to your interior's design. I often try to give all my components one, two, or three colors specifically. This way I can unite all of the components of my system in an artistic way.

Accent lighting can be as simple as getting all of the displays on your source units, gauge clusters, and dash computers to glow in the same color—or it can be as complex as adding LED lights behind interior trim pieces.

Radio Shack is a great place to buy LED lighting. You may also think about using electro-luminescence (EL) lighting. This type of lighting is more expensive, but it offers the most consistent light with the most reliability and flexibility for installation. Razor Lite (www.razor-lite.com) is a great source for EL lighting.

A little accent lighting can go a long way toward drawing attention to your system. This picture shows EL (electroluminescent) lighting that has power lighting it up in the hatch of a Civic. This lighting really stands out in dark spaces.

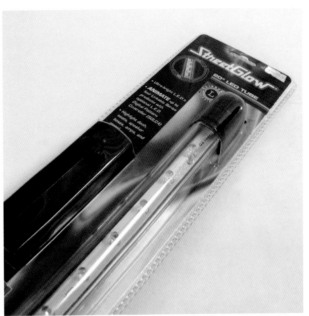

StreetGlow makes an LED tube light that mounts to the vehicle's undercarriage for exterior accent lighting. This is similar to neon but has a brighter glow. It looks really cool at night when you're driving down the road. However, you have to be careful not to use colors that will make you look like an emergency vehicle, such as blue, white, or red. Check with your state to see what neon colors are legal for the exterior of your car.

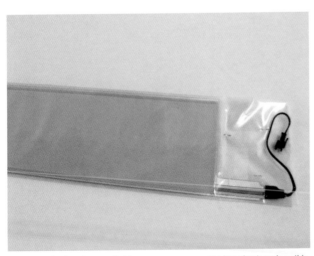

EL lighting that is not connected to a power source comes in a sheet, such as this one made by Razor Lite. The sheets can be cut with scissors into whatever shape you desire. This is really great to place behind cut outs, especially if you want to place it behind the cut out of a name or logo.

Neons and LED lighting can be controlled by remotes. This controller is made by StreetGlow and is made for Neon/LED control. It enables you to change the color of the lighting installed on your undercarriage, as well as control the light to make different patterns for special effect.

LED lighting underneath a car, such as the lighting underneath this show car, can set a mood at a show and really draw a crowd. Plus, it is a lot of fun to drive at night with your car lit up.

LED lighting is also a great highlight on the interior of a car, such as seen on this door panel. At night, this lighting can highlight the custom work in a car.

Neons are another popular choice for accent lighting. StreetGlow is a great manufacturer for neons because their products last a really long time and are very affordable. You can purchase StreetGlow products at Advanced Auto Parts stores. The company makes neon products that can mount to your undercarriage in the form of tube lights, and they also make interior accents lights.

COMPUTERS AND THE INTERNET

Another cool new wave of technology growing in popularity with mobile electronics enthusiasts is integration of laptops and other computing devices into car systems. Adding a laptop to a car can be important, due to the amount of next generation communication tools that are available with the Internet, such as Web TV, Web search, e-mail, and social networking. Also, you can now control your home's security system and other electronics by Internet connections when you use special software.

So imagine if on your way home from work while you're sitting in traffic, you are able to log into the IP address of your stove and begin preheating your oven for dinner. You could even pull up in front of your house and disarm your home security system from your car, while still being able to monitor high-priority work e-mails.

Most cellular phone companies offer wireless Internet cards that are USB 2.0 compatible, which gives you mobile Internet access even when you're in your car. You can also buy a DC-to-AC power converter so that you can plug your laptop into your car and avoid having to run off of its battery. (More on this in Chapter 6.)

If you want to take car PC technology to a higher level, you can even build a PC into your car's dash. To do this, you'll need to buy a computer case and then find a place to install it in your car that will not have too many vibrations (as this can damage the motherboard). You will also need a power supply designed to run off of 12V DC, a motherboard and video card with composite video output, a video screen, a hard drive, and a touch mouse. The audio for the PC can be integrated into the car's stereo system. To help you shop for these parts, try visiting www.mp3car. com. The hard drives sold on this Web site are designed to protect against moisture, humidity, voltage and power surges, and temperature changes. I highly recommend buying PC components from them as they are designed to be inside a car.

It is important to remember that using a laptop while driving can be very distracting for the driver, so if you decide to incorporate this type of technology into your car, be sure to comply with all recommended safety standards.

Some car stereo manufacturers create source units that accept a wide variety of media formats. These source units make it possible for you to play more than just music from them, but also access digital photos, movies, mp3 files, or any other kind of media files. Usually these source units also have a display screen so that you can view these files. Sony makes a model called a MEX1HD that has an integrated memory stick port and a USB connection. This allows you to import mp3 files directly to the unit's internal hard disk from a memory stick. By making your digital media files mobile, you can be sure to have them on hand anywhere you go.

On the Sony MEX1HD, just hit the open button on the face to access the data ports behind the face, as shown in this picture.

As you can see, the memory stick slot is on the left, and the USB 2.0 connection is on the right.

On this Sony XAV-A1 source unit, the yellow 1/8-inch minijack can be used for any external audio and video source to display on the 7-inch touch screen, such as a video camera, laptop, iPod with video, or digital camera.

REMOTE KEYLESS ENTRY

Mobile electronics can include many different types of accessories with various functionalities. For example, a really cool accessory is remote keyless entry. If you have a car that didn't come standard with this feature, then you know what I am talking about. Having to manually lock and unlock your car is a pain in the neck. But for as little as $135, you can add remote keyless entry. To do this, you must buy a $35 door-lock actuator to control the manual-lock function in the car doors. Then you wire a $100 remote keyless entry kit to the actuator to make the locks work by remote control. If you do the labor yourself, you can really get this project done for a small cost. I suggest buying the Spall all-in-one kit that provides both the remote's keyless kit and the actuator. You should plan for this project to take about a day to get done.

Now if you already have power locks in your car but you don't have remote control keyless entry, then Micro makes a great remote keyless entry kit for about $60. This project should only take you around two hours.

BACK-UP CAMERAS

A new security feature that many car manufacturers are promoting is the car back-up camera. This small camera installed on the back bumper of a car can act as a rear-view mirror but with much greater visibility. It also shows drivers their blind spots better.

You can add a back-up camera for less than $100, and they are outstanding if you already have a flip-out video screen in the dash of your car. You might look for a Visor View, which is a military-grade camera. They are easy to install and have great video resolution. You can get these installed in about two hours.

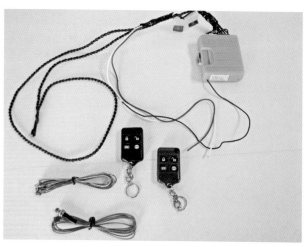

This is a Micro keyless entry system with two remote controls. It should be wired to the power and ground wires at the car's ignition so that it has constant power. The door trigger wires must wire into the factory power-door lock relays, to lock and unlock the doors.

This is a General Motors factory door lock actuator. If you want to get in and out of a vehicle by remote control, you can purchase a keyless entry system and wire it into this door actuator.

A back-up camera can be a great addition to a vehicle for added safety and driving visibility. To install a back-up camera, you should first have your video system installed in the car, connected, and powered. Connect the video cable from the back-up camera to your video LCD monitor. Next, move the camera around to the rear of the car until it shows the desired position on the screen. Once you have found your preferred position, you can mount the camera to the rear bumper of the car. Visor View makes many different back-up cameras, including ones that integrate into your license plate holder, making the camera less noticeable on your car.

GAUGES

There are many types of gauges you can add to your car as an electronic accessory. There are the standard ones that read fuel pressure, air fuel ratio, or water temperature, and other gauges that monitor engine performance. You can also add gauges that monitor your audio system's performance, such as a voltage gauge, a current gauge, or an amplifier temperature gauge. These gauges can be molded into any area of your car's dash, windshield, or center console. You can buy them so that the lighting matches the other interior lights in your car, or you can adjust your factory gauge lights to match your aftermarket gauges. I recommend gauges made by Dakota Digital because they are affordable, have the best looking displays, and have many colors to choose from.

Mobile electronic accessories add the "wow" factor to a car. They are the devices that make people want to go for a ride in your car so they can covet the cool toys. And that's okay—if you work hard to create a cool playground, then by all means, play in it.

Dakota Digital makes voltage gauges, such as the one in this picture. It has brushed aluminum and a blue LED display. They are really easy to wire, and they only have two wires, a positive and a negative. Just be sure to wire the positive wire on a switch so that you can turn the gauge on and off.

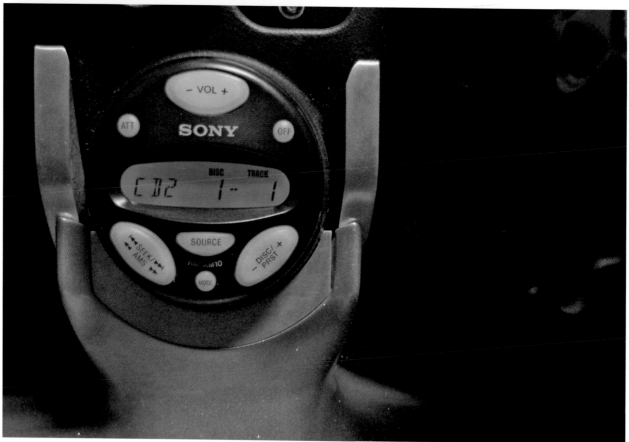

Don't forget that all of these mobile electronic accessories often come with remote controls to operate them. On the top-left is a picture of a Sony Joystick Commander, which belongs to a CDX-CD90 (Sony) source unit. To the right of that is a Sony handheld wireless remote control that controls a touch-screen source unit. The bottom picture is a Sony Remote Commander, which comes with its own LED display and works with most Sony source units. This one is handy because you can control all the functions of a source unit from a remote location.

The toggle switches here were installed in a custom console on a black ABS plastic panel with LED indicator lights. These can be used to turn on amplifiers, lighting, or gauges.

This is an LED status indicator for an alarm system. All alarms come with these, and the best place to install them is in one of the plastic knock out panels in the dash area (found in most cars).

Chapter 3
Planning a Mobile Audio/Video System

Proper planning of any project will determine the success of your final outcome. Installing sensitive electronic components inside tight spaces in a car—which has to both drive down the road and comfortably seat passengers—can be a big challenge. Even for some of the most talented installers, problems and obstacles can arise when you least expect them. Because of this, I cannot stress enough the importance of thinking, planning, and researching every aspect of your design before tackling a mobile electronic system.

Let me share with you an example of poor project planning. When I was in high school, I put a system into my truck. I was pretty proud of the work I did on that truck, but I was also very young and I had a lot to learn about planning a system. When I put the system together, I wired my tweeters directly to the amp. I could crank the system and play some wild bass in that thing.

One night I took my truck up to the ice skating rink to show it off. I felt like the coolest cat in town, with a crowd of people around me wanting to hear my truck. I put in a test tone (bass) CD and cranked the volume up. The truck was vibrating like crazy, and people were impressed. Suddenly, my tweeters started to wail at me like banshees, and the next thing I knew there was a crack that sounded like a bullet splitting through the air. My truck shut off completely. Once people got over that first stunned moment, the laughter broke out. Obviously, I was not the cool guy they had first thought I was. I was quite humbled at that moment.

It turned out that I was very lucky that night. Had I known better, I would have planned my wiring before I put that system together. Then I would have known that tweeters must be wired to a crossover before they can be connected to an amp. By wiring my tweeters directly to the amp, I could have actually blown up the whole truck.

Measure Here

The first thing you need to do before upgrading your source unit is to measure the dash space to see if it is a DIN, DIN and a half, or double DIN. This is a factory source unit in a DIN and a half size. You would want to measure the plastic face of the source unit—the area pointed out in black—to determine the DIN size, and then buy the correct kit.

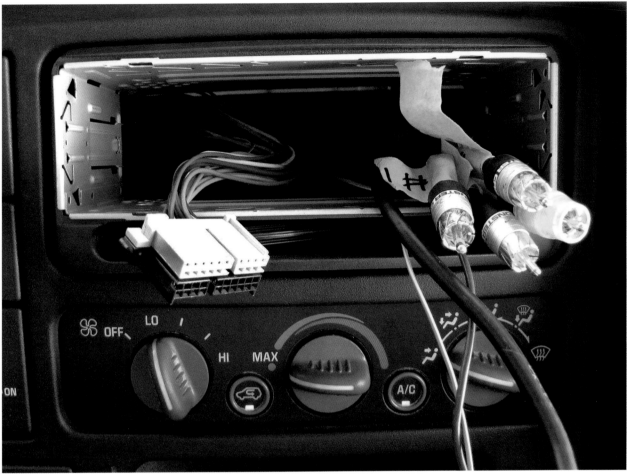

This is an example of a Molex plug behind a factory source unit. This is where you will connect your aftermarket wiring harness.

UPGRADING A FACTORY SYSTEM

Many people choose to upgrade their factory car system rather than spend the money to build an entire custom system. This can be a great decision, especially if you want to be able to sell the car a few years down the road. It doesn't cost a lot of money to make a few quality upgrades to a factory system, and usually you can always take the parts back out of the car and save them for the next vehicle you purchase. Just make sure you remember to save the factory parts in case you ever need to put them back in when you sell the car.

Upgrading your factory audio and video system can be very rewarding with just a new source unit and some component speakers. When you plan to upgrade your factory source unit, you must do the following things:

1. Measure your dash opening to determine if you have a single DIN, double DIN, or DIN and a half. Next you need to purchase a source unit that will fit into that space.
2. You must also buy an aftermarket dash kit. You will need one of these for every single source unit you will ever install. These are readily available at most electronic retailers and car stereo shops. They are made out of ABS plastic, which can be painted to match your interior color. For the most part, they are easy to install. They snap together and mount into your dash where your source unit mounts. Most kits can be purchased for less than $20.

- Keep in mind that kits are specific to your car, and they will accommodate a single, double, or DIN and a half source unit.
- When purchasing your kit, notice that it can be installed using a variety of methods that vary, whether you are installing a single DIN, DIN and a half, or double DIN head unit. (Please refer to Chapter 1 for DIN sizes and photos.)
- If you are installing your kit for a double DIN source unit, you install it without the black storage pocket (which comes with it); if you are installing a single DIN source unit, you need to use the black pocket as a spacer beneath the unit. Likewise, if you're installing for a DIN and a half, the kit will come with a half of a black pocket spacer.

3. You also need to purchase a wiring harness for your specific vehicle. It has a Molex plug on one side, and the other side has color-coded wires. The Molex plug connects into the factory wiring harness behind the head unit; the color-coded wires connect to the corresponding (usually) wiring provided with your replacement source unit. (This will all be covered in detail in Chapter 5.)

• You can get a wiring harness at any electronic or car stereo retailer, usually between $12 and $15. Brand does not really matter with wiring harnesses, but Metra makes a reliable, affordable one.

• **Note:** DO NOT EVER cut the wiring connected to the Molex plug located in the dashboard. This could cause your car to catch fire, among other big troubles!

4. The final thing you need to look for when adding a new source unit is whether or not your car needs an antenna adapter for the new source unit. Source units have a standard sized antenna plug on the back. The antenna wiring of some cars will fit this size, while other cars will use a connector that is too small. If the car has too small of an antenna plug, then you'll need an adapter to increase the size so that it can plug into the aftermarket source unit.

Most cars come with four speakers: a pair in the front doors and a pair in the back seat area. Replacing these four speakers and wiring them to a high-power aftermarket source unit with an iPod connection will give you an extremely enjoyable sound system for less than $500. To improve upon your factory speakers, simply replace them with aftermarket speakers so that they have more power—allowing you to play your music louder with less distortion.

PLANNING FOR A UNIQUE SYSTEM

When you want to design a more complex, intricate system for your car, you need to plan a few additional things before you get started. What is the goal of the finished project? What are you trying to accomplish with your system? You may be trying to have a jamming system for your daily driver, or you may be creating a competition car. Either way, by having clear goals about what you want to accomplish and how you want the final product to look, you will be more prepared to plan your system.

Another thing you need to consider when planning your system is the vehicle that you are going to use for your project. What kind of usable space are you going to have for your system? This is important, because knowing your space parameters will help you decide what products you'll be able to fit into the car. As you begin to shop for your electronics, try to keep the space parameters in mind. This is particularly important with choosing your subwoofer, because a 12-inch subwoofer requires more cubic air space than a 10-inch subwoofer. If you want a low-frequency response with a boost, you may want a ported enclosure for your subwoofer. Ported subwoofers require a larger enclosure size than the same woofer in a sealed enclosure.

Next you need to start thinking about the components you want to incorporate into the system. What kind of speakers do you want to mount? What class of amplifier will you be using? It's a great practice to make a list of the components you want to use and then visualize how they will work together. Some speakers and amplifiers complement each other better than others.

Consider the type of speakers you're going to use, how they will mount, and the location where they will mount. If you are planning a unique system, you'll probably want to put the midrange or upfront speakers into the kick panel area within a solid enclosure, instead of mounting them in the door panel. Speakers also require a power source, so you need to plan to for the power requirements the speakers need in order to play loud and clear.

If you want your system to play loud, then buy high-powered amplifiers. Big amplifiers will give you big power, but that is only half of the equation. You also have to give a quality signal to the amplifiers in the form of voltage from

---

TIP!

Here are a few examples of unique system components that go well together in both speed and amplitude:

• 2 x 6.5-inch speakers and 1 x 10-inch sub
• 4 x 6.5-inch speakers and 2 x 12-inch sub
• 6 x 6.5-inch speakers and 2 x 15-inch subs or 4 x 12-inch subs

---

Basic Human Hearing Is Ten Octaves Of Music

20Hz	120Hz	400Hz	5KHZ	20KHZ
Sub-Bass Frequency	Mid-Bass Frequency	Midrange Frequency		High-Range Frequency
Subwoofers	Mid-Bass Drivers	Midrange Drivers		Tweeters

This chart shows a frequency response of 20 Hertz to 20 Kilohertz, which is the normal range of human hearing. It also shows what frequencies belong with which speakers. For example, 400 Hertz to 5 kilohertz is mid-range frequency and should be set to mid-range speakers.

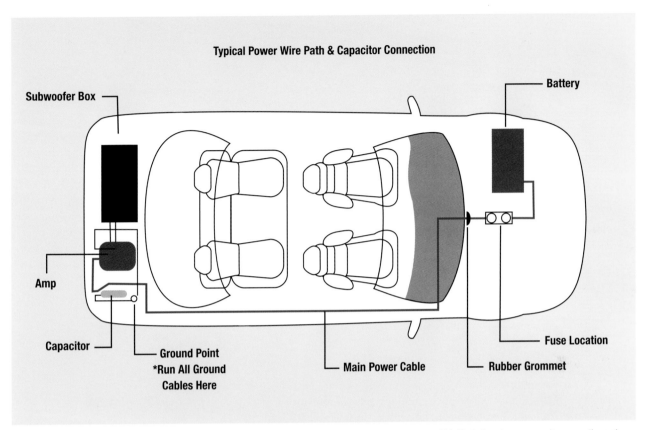

Typical Power Wire Path & Capacitor Connection

Subwoofer Box

Battery

Amp

Capacitor

Ground Point
*Run All Ground
Cables Here

Main Power Cable

Rubber Grommet

Fuse Location

When planning a system, you need to draw a diagram of your wiring schematic before you begin installing your system. This illustration shows a capacitor connection and fusing connection points. By planning these wiring paths, you can visualize how your system electronics will lay out and how they'll be installed into your car.

an RCA cable. Source units have RCA outputs that send voltage through the RCA cable. A quality source unit will send 4 volts of preout voltage to an amplifier. If the source unit signal is less than 2 volts, it can and should be amplified by a line-driver (which is an audio component). These are relatively inexpensive (around $50) and should be wired in-line between the source unit and the amplifier. This tiny device amplifies the signal voltage before the amplifier gets hold of it. This process increases the signal-to-noise (SNR) ratio, which means you are amplifying the signal in hopes of prohibiting noise from getting into the RCA cables. Most RCA cables have 75 ohm impedance and act as a giant RF antenna. An RF antenna is like an FM radio antenna—it attracts noise. The longer the RCA cable, the longer in essence the "antenna" and the more noise you will attract. Thus the line-driver amplifies the signal and reduces the chance of noise in the cable.

Another very important thing to think about when planning your system is that you must use heavy-gauge speaker wire. Any unique or high-power sound system should use 12-gauge speaker wire or larger.

Subwoofers are a major part of your unique system. Try to plan for whether or not your subwoofers will keep up with the amplitude or volume level of your front speakers

and keep up in speed and timing with the rest of the system. There's nothing worse than listening to a system that has fast, articulate-sounding midrange and a slow, sloppy subwoofer!

Electrical currents requirements are another item to plan for in a unique system. You have to calculate how much output power (watts) your system will have and then calculate the size of power wire you'll need in order to send enough current through to your amp, as well as what amperage of fusing you need on the line.

Here are some other things to consider when planning for your unique system:

1. Will you need sound-deadening materials?
2. What kind of security will you use to protect your system?

COOL TIP!

Try to choose amplifiers that have removable heat sinks that can be painted to match the interior color of your car. You could also sandblast (which is a method for smoothing or finishing a surface by using sand blasted through a gun) or chrome the heat sinks to match your design theme.

SAMPLE BUDGET

Total Amount of Budget: $5,000

Car: 2004 Honda Civic

Project Goal: A simulated show car appearance in a daily-driver car, including necessary accessories for road trips

1. Source unit with 4-volt preout and iPod connectivity and Bluetooth; ability to upgrade to satellite radio = $500
2. Two JL Audio 10-inch subwoofers (10W3V6) = $500
3. Sony XPLOD XM-SD22X amplifier for front of car = $200
4. Sony XPLOD XM-SD61X amplifier for rear of the car/subwoofers (class D digital) = $400
5. Two sets of Sony XPLOD component speaker sets (XS-D170SI) = $400
6. TomTom GPS navigation unit = $200
7. 8G iPod Nano = $100
8. Sirius satellite radio upgrade = $100
9. Audio Control Matrix line-driver/crossover = $300
10. Capacitor = $100
11. 4-gauge power and ground wires (25 feet total) = $150
12. RCA cables = $50
13. Fuses = $50
14. Wiring harness, kit, antenna adapter = $40
15. Heat shrink and loom = $25
16. 100 feet of speaker wire = $60
17. Viper car alarm with starter kill and auto page = $300 (This will include installation.)**
18. Video kit with DVD player, two pairs of wireless headphones, and 10-inch overhead monitor = $550
19. Ring terminals, spade terminals, butt splices, fork terminals = $50
20. Dynamat (and other sound dampers) = $300
21. Meguiar's car show cleaning kit: glass cleaner, Speed Detailer, wheel cleaner and wheel polish, Super Protectant, Quick Interior Detailer, microfiber cloths, and chamois = $115
22. Dacron (pillow stuffing for subwoofer enclosures) = $10
23. Materials for fixture(s) and enclosure(s) (wood, carpet, glue, screws, nuts, bolts, and so on) = $500

TOTAL = $5,000

**It is recommended to have a professional install an alarm for insurance purposes.

Here is a sample budget for a mobile electronic project. Make sure you plan your system before you begin, so that you can make sure that the whole project is feasible with your estimated budget.

INTEGRATING YOUR AUDIO/VIDEO COMPONENTS INTO YOUR VEHICLE

When purchasing your audio/video components, it is essential to note their specs, connections, price, and design. This is often overlooked. It's important to choose components that (1) match the existing interior of your car or other product, and (2) are modifiable to match your system design and theme.

Also important are the layout and mounting of the components. I suggest doing this in an area in which the products will most easily integrate or look natural with the rest of your interior. If this isn't possible, then mount them in an area that is easily customizable so that they blend in with the car.

When mounting and placing your speakers, aim to use the same design theme throughout your car. For example, if your kick panel enclosures are wrapped in vinyl, then all of the speakers in the rest of your car should be in enclosures wrapped in vinyl. If you have visible door speakers wrapped in suede, then try to highlight your other speaker installations in the car in the same color suede. If your dash has black plastic around the instruments, it only makes sense to have back plastic around your source unit as well. These are the sorts of things that make all the difference in your design and make your car different from everyone else's.

Finally, try to choose your design theme before you begin your project, and then stick to the theme from front to back as you work on the installation. This will help your car have a consistent, sleek, and professional look.

This picture shows the results of a poorly planned and executed stereo system. The ground cable coming off the negative battery post is connected to the capacitor and then fused to run back to the amplifiers. This is a recipe to blow up a car. They should have run the power cable off the positive post on the battery, and the capacitor should have been in the car next to the amplifiers.

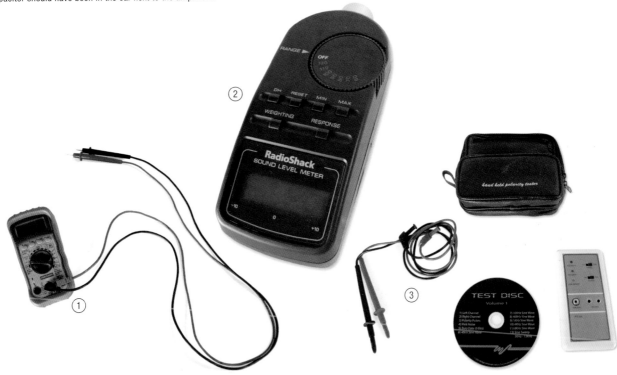

In planning a proper audio/video system, there are some must-have tools that will assist you with your build. Here are three of those must-have tools. 1. A multimeter measures voltage, impedance, continuity, and current in your wires or electronic components. 2. A digital sound-pressure level meter measures volume. 3. A speaker-phase detector determines if your speakers are in phase and wired correctly.

SHOPPING LISTS:

Composite Materials

- Fiberglass mat (0.75 and 1.5 ounce)
- Polyester fiberglass resin
- Paintbrushes (2-inch, wooden)
- Mixing cups and lids
- Strainers
- Stir sticks
- Acetone
- Latex gloves
- A methyl ethyl ketone (MEK) dispenser
- Gel coat
- PVA (polyvinylalcohol) mold release
- A MIL gauge (MIL is a measurement of circular area)
- Plaster of Paris
- Shellac
- Paraffin wax
- Sandpaper
- Body filler
- Duraglas body filler
- Kitty hair
- Cabosil silica
- Polyester primer surfacer
- 2K Primer
- Guide coat
- Texture coat
- Glazing compound
- Razor blades
- Brad nails
- CA glue (cyanoacrylates; can be found at hobby stores)
- ABS plastic
- 6061 Aluminum
- ABS Rod Solid
- ABS Rod Tube
- Aluminum Rod Solid
- Aluminum Rod Tube
- Cold Rolled Steel (Mild)
- Acrylic Rod
- Acrylic Sheet
- Laminate
- Laminate glue
- Lacquer thinner
- Metal mesh
- Apple ply veneer
- Grille cloth
- Vinyl
- Carpet
- Nut rivets
- Bolts
- Screws
- Threaded rod
- Angle iron
- Double-sided tape
- Electrical tape
- 3M 233+ green masking tape
- Visquine
- Brown paper
- Silicone
- Polyurethane adhesive
- Great Stuff Expanding Foam
- Black spray paint
- Hot glue
- Goop hand cleaner
- SEM 20-second and 3-minute epoxy

Wood Materials

- MDF (⅛-inch, ¼-inch, ½-inch, ¾-inch, 1-inch)

Electrical Supplies

- Fork terminals
- Ring terminals
- Spade terminals
- Butt splices
- T-taps
- Relays
- Fuses
- Solder
- Crimp caps
- Zip ties
- Heat shrink
- Loom
- TEK Flex
- Primary wire
- Speaker wire
- Power wire
- Battery terminals

Basic Tools

- Jigsaw
- Trash can with lid
- Inverted router
- Handheld router
- Skil saw
- Chop saw
- Table saw
- Heat gun
- HVLP (high-volume, low-pressure) gun
- Glue gun
- Canister gun
- Drills
- Drill bits
- Level
- Square
- Air compressor
- Rulers
- Multimeter
- Hole saws
- Screwdrivers
- Wrenches
- Shop vac
- Scribe
- Razor knife
- Angle finder
- Pencil sharpener
- Tap and die set
- Crimpers
- Cutters
- Pliers
- Hammer
- Rubber mallet

You can see the bird's nest of wires with two different amplifiers and a subwoofer box that is not bolted down to the car. This is a fire waiting to happen. Always lay out your wires neatly and then zip tie them together. Match your amplifiers as much as possible, and hide your system with a hatch-cover to protect against theft. Always bolt down a subwoofer box, so it doesn't move around when you're driving—this is a safety precaution!

This is an example of a properly planned and designed system, in which the hatch cover protects the installation. All of the wiring is cleanly organized and hidden, and the system is installed where the back seat still functions properly. All of the equipment is bolted to the car to ensure passenger safety. The opening in the center of the hatch cover creates an easy-access point for wiring and fuse blocks. When planning your system, make sure that you leave openings that will make it easy for you to get back into the wiring at a later point, in case repairs or adjustments need to be made.

PLANNING YOUR WORK SPACE

Every installer needs a space where they can do their work. This space needs to take into consideration elements, such as the weather and general functionality. You don't want to work in a driveway where the wind, rain, sun, or snow can deter your work. You also don't want to be too far away from where you keep your tools, because you will constantly need to be grabbing a different tool or piece of hardware to work on your project. Try to think about where you can park the vehicle for a while where it will not have to move.

It's also important to have a workbench where you can assemble products. Most equipment that you use for the car will need to be unpacked, assembled, or modified in some way. A work bench can give you counter space for doing this type of assembly.

A tool cart is also helpful, because it allows you to move your tools around with you more easily. You can make your own tool cart pretty easily and inexpensively, but if you don't want to spend the time you can buy them, too. Just make sure it has upper and lower storage for various sized items and wheels so you can move it around easily.

Probably the most important part of your workspace is lighting. I don't think it's possible to have too much lighting. The overall work environment needs to be lit

Shown here are three different electrician's pliers. Each has a specific function. On the left are wire strippers; in the center is a compression crimper (crimps butt splices, fork terminals, and ring terminals without damaging the plastic insulation on the connector); at right are wire cutters with a very sharp edge. These are great for cutting the excess from a zip tie after you have secured wires with them.

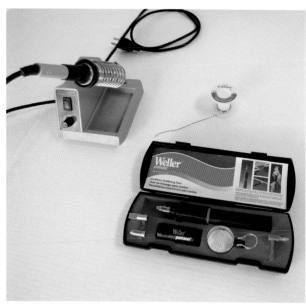

For best results, use soldering instruments to connect wire-ends. The soldering iron melts the solder material and joins both wires together. Once the solder cools, the wires virtually become one. At left is a temperature adjustable soldering station; at right is a cordless soldering tool.

Heat guns are most commonly used to activate heat shrink. They are also used to remove old Dynamat or sound-deadening material. You can even use one—lightly, from a distance—to clean up frayed carpet edges. Heat guns can even be handy for loosening rusted bolts.

A hood prop can be an invaluable tool. This model is a spring-loaded aluminum rod that has padded, rubber bushings at the top and bottom. It's really useful if you're working in the back of a hatch car with gas shocks that normally hold up the hatch. This tool can be used in case you have to take the gas shocks off the car to do any kind of work. The tool props the door up while the shocks are off. It costs around $20 from Mac Tools.

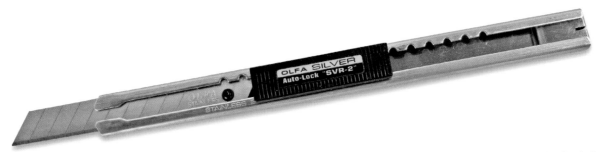

This is a retractable razor knife made by Olfa. It's extremely thin and lightweight and has a razor sharp edge from a stainless-steel blade. It is best used for trimming plastic, cutting carpet or cardboard, and trimming visquine. You can buy one at Lowe's for around $5.

This is an upholstery finish stapler. If you aren't one for sewing, this is a great tool. It is also the one and only tool you should use to wrap carpet on a speaker box or to wrap vinyl or leather around a door panel.

I like to use a Bosch jigsaw, which was my very first major tool purchase. For more than three years, I built many competition systems with a $20 Craftsman jigsaw. I finally decided that I needed a tool that would cut straighter and with more power, so I upgraded to this Bosch jigsaw, which has orbital cutting action. Orbital cutting action means that the blade pushes forward and cuts at the same time, almost in a pendulum swing. This means that it pulls back just as the cutting debris is just starting to build up. You wind up with a straighter, faster, cleaner cut. I have had this Bosch jigsaw for more than 10 years, and it still works like new.

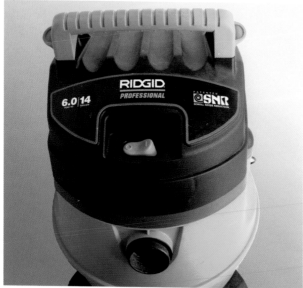

Don't forget to buy a shop vacuum. This is going to be your best friend during a car build, and you might want to think about having more than one. They are not only good for cleaning up after yourself, but you can use them to clear away sawdust while you are cutting or routing wood. In my opinion, the only brand worth buying is Ridgid.

This is the Bluepoint LED light kit that comes with a hanging attachment and a spring pivot attachment, as well as a rechargeable battery. This tool is useful under dashboards, radios, or any small working areas where you need a lot of light. Also in this picture at the bottom is the Surefire flashlight, also great for getting a lot of light into small spaces. You can buy these flashlights at Lowe's for $69.

up with multiple fluorescent work lights—preferably the ones that are 8 feet long. I use the cool white bulbs for purer lighting.

If you're working inside of a car, you'll need a 500-watt halogen work light shining in through the opposite side of your workspace in the opposite door. On the side of the car where you are working, you'll need a hand-held fluorescent work light with a hook to place it in the immediate vicinity where you're working. These are especially useful for working under the dash, hood, or undercarriage of the car.

Blue Point makes a light kit for around $60 that comes with several different attachments. It's easy to get creative with positioning the light however you need it, and it has a rechargeable battery.

Surefire also makes a handy little flashlight that is only 4 inches long. It is 80 lumens, which is the brightness of two Maglites put together. However, because of its petite size, you can get into small spaces. This flashlight is often used on rifles and other long-range guns for sharp shooting, because the light is extremely white and clean and prevents shadows from blocking your view.

When you set up your work space, you should get an air compressor. Air tools are extremely flexible and more efficient than electric tools. They are also very affordable. Grinders and cutters have a host of different tips, attachments, and blades that you can use for any type of fabrication that you're doing. Sanding with an air tool can really cut down on the amount of time you spend sanding fiberglass. Compressed air is also very useful. It can be used for cleaning dust and debris out of your shop, car, and clothes. It's also good for blowing sanding dust off the part you are fabricating before you apply any layers of body filler. You can also use the air my clothes, hair, and shoes before leaving the shop for the day.

This keeps the dust out of my cars and my house after a long day's work.

If you get an air compressor, try to get one with a tank that holds at least 30 gallons to start out with. The more you can afford, the better, because you will go through the air quickly. Also try to get no less than 2 running horsepower.

When you add an air compressor to your shop space, you will need a place to plug it in. A 30-gallon air compressor runs off a normal 110-volt outlet. However, if you get a 60-gallon compressor, you will need a 220-volt outlet at 30 amps. This is the same as a plug you would use for a laundry dryer.

Most handheld power tools need a 15-amp receptacle minimum for prolonged use. These 15-amp receptacles should have a number 12 wire running up to them. You will also need several extension cords that have a number 12 wire inside of them, and they should be rated for 15 amps. This information is listed on the packaging for any extension cord.

Table saws and routers are must-have tools. One of the most popular table saw choices is the Unisaw by Delta, and it comes in 3-horsepower and 5-horsepower models. It also has both left-tilting and right-tilting blade versions. I prefer the left-tilting blade because the right side of the saw is always the fence. If you're cutting a bevel on a piece of wood, then a right-tilting blade is more likely to bind the piece of wood to the fence and cause kick back and/or injury. Both of these versions of table saws require a 220-volt power outlet for their plug. Sears has a Craftsman Professional hybrid table saw. It can cut through heavy, thick wood, but it only requires you to use a 110-volt power outlet.

Routers will usually need between 15 and 20 amps of power out of their outlets. For all your electrical needs, I recommend having a state-licensed electrician come in to your shop space to check all of your outlets and wire

Snap-On makes a ratchet that has a rotating head, making it perfect for tight, small places. It can also be used in an instant as a nut driver if needed.

any new receptacles. They can help make sure that all your power is sufficient for each of your tools and that using more than one tool at a time will not short out your power. It's also a good idea to have them help install your in-ceiling shop lighting, too. The last thing you ever want to have is sloppy electrical work around your valued car investment. Give yourself the peace of mind in knowing that the electrical is up to code so you don't have to worry about burning down your shop.

I used to do all of my shop work out of my small, two-car garage. The garage was also the home of my washer and dryer. I had my professional-grade 60-gallon air compressor set up next to the dryer so that they could share the 220-volt outlet. Every time I used my dryer, I unplugged my compressor and vice versa. Over time, this plugging and unplugging wore down the outlet, and the dryer plug and the wiring behind the outlet came loose. It was touching the metal junction box behind the outlet and began to spark and burn the wires. Had I not noticed this issue, it would have continued to a point where it could have burned down my whole house.

The moral of the story is to not let your power tools and other equipment share plugs. If there are not enough receptacles available, then have more wired in by an electrician.

Take it from me when I say that you will thank yourself for this safety precaution.

Finally, you need to consider where in your work space you will do your woodworking. There is so much saw dust that comes from woodworking that it can infiltrate your entire working space. I highly recommend that you set up all of your woodworking in a separate, contained area. If you can, try to set up a wood shop with router tables, workbenches, and a space for lumber storage. This way, your garage space can be kept clean from dust and can be dedicated to your wiring and installation projects.

Shopping Lists

When doing an installation for yourself, you need a wide variety of materials as well as access to some basic tools and small parts to ensure that your installation can be completed. You'll also need a secure place to work on your vehicle that is covered and out of the weather and that also has proper lighting.

As you work with composite and wood materials, always wear a small-particle cartridge respirator and impact-resistant eye protection, and work in a well-ventilated area. No matter how silly these safety devices may make you look, it is far less humiliating than having to go to the hospital for an avoidable injury.

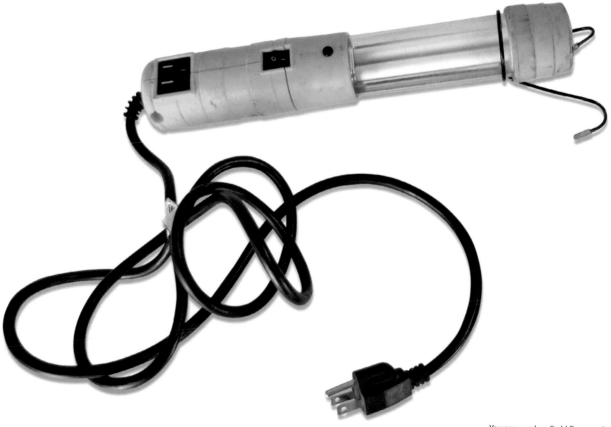

You can use handheld florescent lights with hanging hooks to light up an entire floorboard. Look for these at Sears for around $20.

This florescent light is made by Snap-On and has spring-loaded hooks, which hold the light up to the underside of your hood or roof of the car. It's perfect for lighting up the entire engine bay or interior of the car. It also has a very long cord that makes it easy to work with.

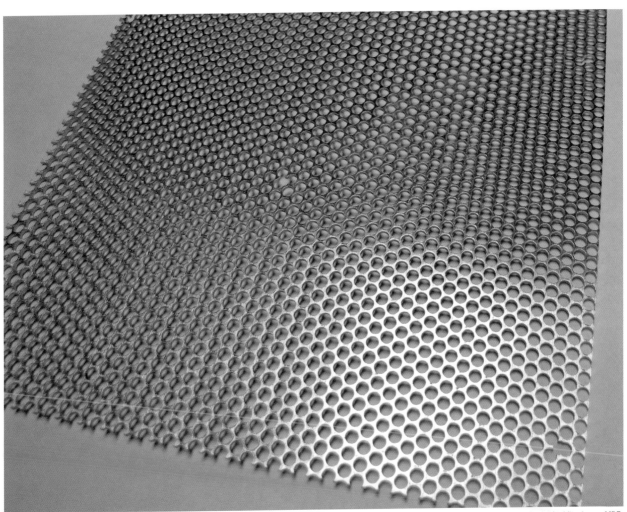

Metal mesh is perfect for making custom speaker grilles or a cover for a subwoofer port. If you buy this in aluminum, it can be cut on your router table with a flush trim bit using an MDF template. If it's made out of steel, you'll have to cut it with a plasma cutter.

A quarter-inch sheet of black ABS plastic can be cut easily with a jigsaw and then routed into a desired shape. This black plastic is great for radio trim rings, speaker grilles, and switch panels.

Here is a list of places where I recommend shopping for your materials:

- Ace Hardware
- Lowe's
- Home Depot
- McMaster-Carr (nuts, bolts, hardware)
- Mobile Solutions-USA (router bits, specialty router tools, car audio install tools)
- U.S. Composites (fiberglass resin)
- FinishMaster (bondo, auto body paint and primer, chemicals, lacquer thinner, acetone)
- Piedmont Plastics (LEXAN and acrylic and ABS plastic)
- Wurth Wood Group (MDF sheets)
- Woodcrafters (wood and woodworking tools)
- Michael's Arts and Crafts
- Mae's Fabrics
- Wal-Mart
- Advance Auto Parts
- Harbor Freight Tools
- Northern Tool
- Best Buy
- Radio Shack
- Sears
- Pep Boys
- NAPA Auto Parts
- Clarity Cable (high-end audio cables)
- Intro Custom Wheels
- Streetglow (neon and LED lighting)
- Meguiar's (car cleaning supplies)
- Klingspor (sandpaper, abrasive materials)
- Discount Tire
- KW Automotive (suspensions)
- Stainless Steel Brakes
- Optima Batteries
- Monster Cable
- Dynamat
- Dakota Digital (voltage, temperature, and current gauges)
- Covercraft (car covers)
- Katzkin Leather (seat covers)
- Razor-Lite (EL interior lighting)
- SPC Performance (camber and castor adjustment kits)
- MagnaFlow Performance Exhaust
- Exedy Globalparts (racing clutch)
- Toyo Tires
- Sparco (seats)
- Beltronics (radar detectors)
- Miller Electric (welding equipment)
- Evercoat (body filler, fiberglass resin, glazing compounds, putty)
- Parts Express

All of these companies have Web sites on the Internet, and most of them offer you the option to order online. Shopping for bulk products online can save a lot of money. For example, no company can beat the bulk price of fiberglass resin from U.S. Composites, and McMaster-Carr has the lowest prices on bulk screws, nuts, bolts, and other miscellaneous hardware.

When I shop for raw materials, regardless of where I live, I find local yards and distributors that sell aluminum, wood,

Plexiglas is made of acrylic. It can be machined or drilled and is very useful for making amp covers. You may also want to use it to make fuse panel covers or battery covers.

plastic, or steel. I like to buy these items from different places. If I find a local steel company with the most competitive prices, I will buy all of my rods, sheets, and flat stock from it.

If I need metal mesh, I tend to ask my steel or aluminum vendors to compete for a price to sell me. You don't have to be a business or contractor to work with these types of companies, and they can save you a lot of money on your materials.

WHAT TO DO WHEN YOU GET AGGRAVATED

If there is anything you should plan for when working on a system, it's getting frustrated. Let's face it, no matter how hard you try, things are going to get muddled up and go wrong, and you're going to get mad. Don't worry; it happens to everyone. There is nothing more aggravating than installing a car stereo system. There will be at least a few times when what you are doing is just not going to work. At the same time, there will be things in your way of making it work.

Here are a few simple tips to get you back on track when you get aggravated:

1. *Stop.* When you get mad, stop what you're doing. Don't force what isn't going to happen.
2. Try to figure out what you can do to make the problem go more smoothly. At lot of times, your environment can be a contributor to the problem. Take some time to figure out what is in your way. Start to get organized in your working space, and this will help your thoughts become more organized, too.
3. Get more lighting. Many problems arise because you can't see well enough, so add some more light to the space.
4. Check the temperature. When you are too hot or too cold, you are uncomfortable, making it hard to focus. Adjust the temperature until you are feeling better.
5. Get something to eat. It's really easy to get carried away on a project and lose track of time. Chances are good that if you're mad, you're also really hungry.
6. Take a power nap. If you've been on an all-night bender trying to finish up, then you are tired. Stop where you are, and take a 10- to 15-minute nap. This will refresh you enough to get your mind back on track.
7. Take a step back. I often like to just step back a few feet and look at what I am doing. I ask myself, "What tools am I using here? Are they working? Are there other ones that would work better for this project? Can I be doing this process differently?"
8. Narrow your tasks. If you're mad, it can often be because you are trying to do more than one thing at a time. Multitasking is never a good idea with an installation project, so narrow what you're doing down to just one task, and this should help ease your frustrations.
9. Google it! When all else fails, get on Google and look for troubleshooting tips. Chances are good that you'll stumble onto a Web forum where other people are discussing the same problem. You can often locate good advice this way.
10. Find a friend. Sometimes all it takes is someone else to bounce your ideas off of. Conversation with another person can really help you get your thoughts together, even if the person only listens.

Chapter 4
Automobile Acoustics and Noise

You probably have experienced a car that is noisier than its sound system as it drives down the road. Let's face it, some cars are just piles of metal that bump, rattle, grind, squeal, and roar. It often seems that you'll never be able to muffle the sounds that stream from a car. However, I assure you that even the most run-down heaps of junk can be softened to sound acoustically correct. The better you make the acoustical presence in your car, the better your sound system is going to perform.

THE NOISE FLOOR INSIDE VEHICLES

The noise floor is the base level of perceived sound that makes its way into your vehicle's interior while you're driving your car. This includes the roar of spinning tires, the sound of the exhaust, or the hum of a working engine. In the car audio industry, this extraneous noise is called the noise floor.

The average noise floor level inside most passenger cars is about 90 decibels. You may have noticed that your ability to hear low-level sounds from your stereo system, Bluetooth, or cell phone can suffer because of your noise floor unless the volume is cranked on your devices.

All audio systems have what is referred to as "dynamic range." Dynamic range is the difference between the sounds in your music at low volume and the sounds of your music at high volumes.

The more ways that you find to reduce your noise floor levels, the better you will be able to hear the low-level sounds transmitted by your stereo system or mobile accessories. By reducing the noise floor, you will be able to hear more of the dynamic range of music and more sound from your system.

SOUND DAMPING BENEFITS AND APPLICATIONS

Stage 1

Let's talk about how you can go about lowering the noise floor. Any kind of vibration in your car (tires, suspension, engine, exhaust) can get into the interior of your car. These vibrations cause the metal panels of the car, both inside and outside, to vibrate and send a nasty sound into your interior. The plastic interior panels inside your car cover up inner sheet metal, which can also vibrate and cause problems.

Your goal is to minimize the vibrations of the metal inside your car. You can do this by using sound damping, by treating the metal to prevent it from vibrating. Think of a tuning fork: Once it vibrates it produces a sound. The same is true for the metal in a car: Once it vibrates it begins to produce sound. The only way to reduce the vibrations on metal surfaces is to wrap them in an alternate material. Think about the tuning fork again: If it's vibrating and you wrap your hand around it, the sound becomes muffled.

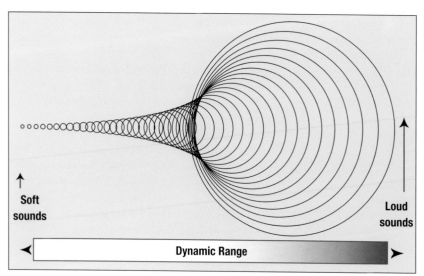

Dynamic range describes the range of musical information that can be heard between the softest and loudest sounds within a track at a constant volume.

Soft sounds

Loud sounds

Dynamic Range

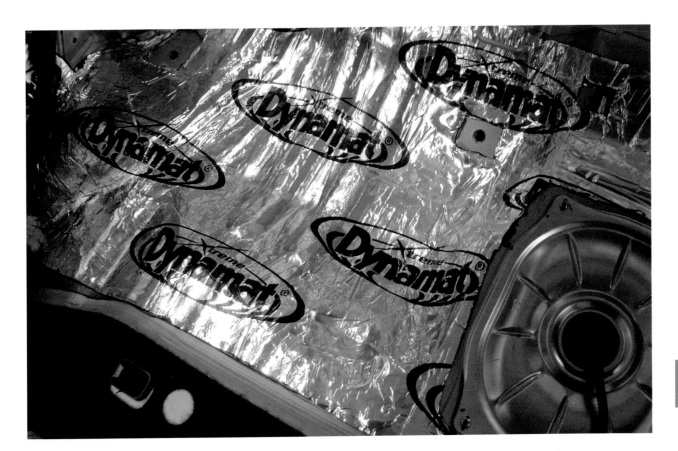

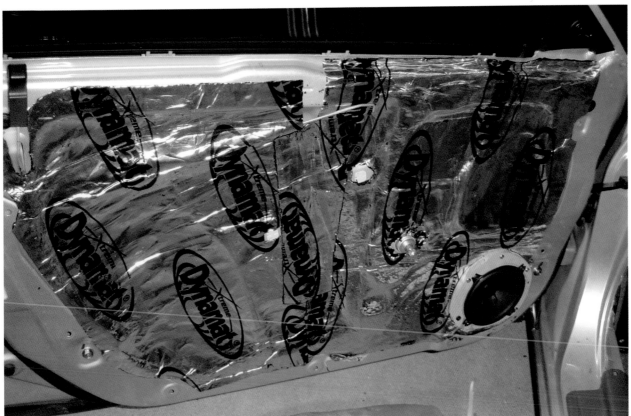

Dynamat is one of the best products to use to damp sound in your car. In these pictures I applied it to the doors and floor of the car.

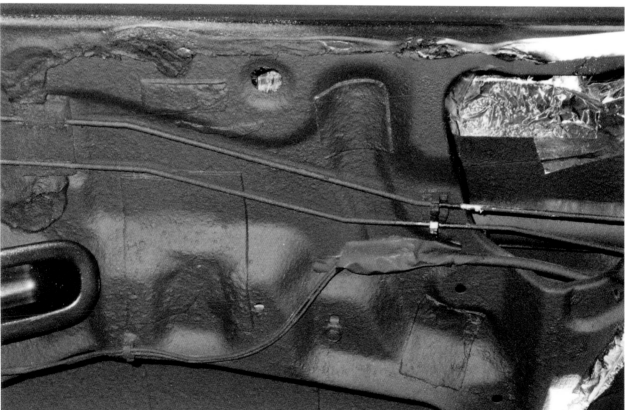

DynaShield is a great product to use on your car doors. These pictures show the car doors before and after the DynaShield was applied.

Sound damping material options can vary greatly. My personal preference is Dynamat. I have had extremely good results with Dynamat over the years because it creates a tomb or bank-vault kind of effect inside a car—making it quiet and vibration-free.

Another sound damping option is spray-in material. This can be done with DynaShield, which is made by Dynamat. Please note that you cannot put Dynamat over DynaShield. Most people use one or the other. I would say that the Dynamat on the door is the better choice. The most optimal scenario would be to apply the Dynamat to the outer door skin and then to the inner door skin. After that mask and prepare the door to spray with DynaShield over the Dynamat to seal everything. DynaShield is an aerosol spray, and the overspray that gets into the air from this product will get on your paint and in the interior of your car. Be sure to mask off your car and any other areas where you are spraying with visquine. You can also get the DynaShield effect by purchasing truck bed liner and rolling it inside your car's interior (while the carpeting is out) with a foam paint roller. Be sure to mask the area off with paper and plastic.

Obviously, the best-case scenario would be 100 percent treatment of the car's metal. If you have the budget, go 2, 3, or even 10 layers thick with your sound-damping materials. IASCA competitors (such as myself) have made the sound barrier up to an inch thick.

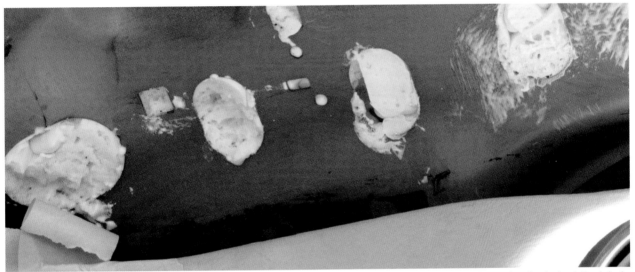

Great Stuff Expanding Foam is the best way to get sound damping into hard-to-reach crevices. This picture shows the A-pillar area of a car, where the sheet metal has holes in it. The spray foam was used to seal the holes in the metal.

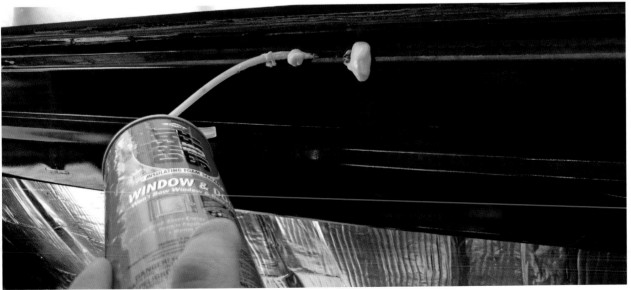

Here, Great Stuff Expanding Foam is applied to a body cavity in the back of a truck's cabin to reduce vibrations and road noise.

Stage 2

The next noise treatment you can do should be performed in addition to Stage 1, not in place of it. In this step, pack any compartments or void spaces between your interior panels and your vehicle's exterior sheet metal full of insulation. If it's an area not prone to getting wet, you can use R30 fiberglass insulation, available at your local hardware store. This is the absolute cheapest way to insulate your car.

If the areas you are filling are prone to getting wet, you'll want to fill them with waterproof foam instead of fiberglass. Try using Great Stuff Expanding Foam, also available at hardware stores. It comes in a can with a blue label marked "Window and Door." Be sure to get the one that is for "minimum expansion" so that you do not buckle the sheet metal when the foam expands. Another really cool trick is to take the same foam and use the included straw attachment to fill the chambers around the A- and B-pillars in your car. Look for the 1-inch-wide openings in the sheet metal in your car where the interior trim panels clip into place. You can push the straw into the clip opening and squeeze the foam into the area until it fills up. Make sure that you leave enough room for expansion.

The next part of this phase is to cover up the flat floorboard areas with a thick layer of squishy foam. Dynamat calls its product DynaLiner, but I like to use their ExtremeLiner first. ExtremeLiner is a thin layer of lead, a thin layer of foam, and a thin layer of rubber sandwiched together. Once the ExtremeLiner is down, place the DynaLiner over it. As an alternative method, you can go to any local fabric store and purchase thin, yellow, ½-inch or 1-inch foam and place it down on top of your damping material and under your carpet. As a word of caution, Dynamat is an incredible product, and I highly recommend it. However, it can be messy, and if you get it on your clothes, it probably won't come out, so I recommend wearing clothes you don't care about getting a little dirtied up.

ExtremeLiner is great for car floorboards; it helps reduce the highway noise in your car.

Wrapping movable door parts in foam helps reduce the rattling sounds they can make when the car is being driven. You can also use this product to wrap your pedals and other moving parts. Or, like in this picture, you can even use this product to wrap your wires to keep them from resonating against the floor. However, if you plan on driving the car too, then wrap movable parts only when it won't affect drivability.

Stage 3

If you're ready to take your acoustical treatments another step forward, then wrap all of the moving pieces inside your door panels (door striker, handle, and latch parts) with an adhesive-backed foam or felt material. (This can be the type of felt fabrics found in a fabric or crafts store, or it can be weather-strip material found in a hardware store.) Place the foam or felt in any panel seams or areas inside the door skin where you can press it into place. This method of sound damping is great for filling in the small places where you can't get the materials from Phase 2 to fit.

Another great technique is to take all of the wiring harnesses inside the car and wrap them in the adhesive-backed foam material (like weather-stripping material) like a candy cane stripe. You can also apply sound-damping material such as Dynamat to the back sides (not seen) of all of your interior plastic panels (door panels, dash parts, console parts). A similar sound-damping option that works well is to stitch foam to the underside of your floor mats.

The roof of your car can be very noisy, so it is critical that you remove your headliner and apply sound damping to its underside. Equally important is to remove your front fenders and apply Dynamat to their backsides in a few layers.

As you can see, your imagination is the only limit to what you can do with sound damping. If you can remove the part from your car, you can apply sound damping to it. You can never apply too much damping, so layer it on until your budget runs out—it's certainly worth every penny. However, you should note that sound damping does weigh down a vehicle. If you will use your car as a daily driver, then you need to calculate how the weight of the damping will affect the drivability of your car. If you use a lot of damping, you will probably need to consider buying better shocks and springs to support the weight of the car.

Let me illustrate this point with a personal example. I added concrete as a sound damper to one of my competition cars. I mixed and poured the concrete into the floors and fender wells. It is a great sound barrier and really seals the system. Because the weight of the audio system in this particular vehicle is already so extremely heavy, the weight of the concrete actually helped to balance the car back out a bit. It drives well, but it is very heavy. I have to use a special suspension system to support the weight of the vehicle. Although the concrete makes the car sound fantastic, I don't recommend this approach except for most custom of competition systems. I say this for two reasons: 1) You are going to seriously affect the drivability of the car; and 2) concrete is permanent. Once you put it in, it's not going to come out. Remember to think of these sorts of things as you determine which types of sound deadening are going to work best for you.

It's a good idea to put sound damping in your headliner, too. Remember that sound can come into your car from anywhere. Just keep in mind that the headliner can be tedious to remove and replace without damaging it, so take this project slowly and carefully.

INTERIOR ACOUSTICS/ABSORPTIVE SURFACE AND REFLECTIONS

Interior acoustics inside most cars are really terrible. If you calculate the cubic air space inside your car (LxWxH), you'll find your car is smaller than most closets. There are usually four or more seats in a car, which are surrounded by glass and hard, shiny surfaces. If your car has leather seats, your interior acoustics can be even worse. You are faced with acoustical challenges because sound waves bounce off of shiny, reflective surfaces.

Sound waves radiate toward a listener in the same fashion that ocean waves roll toward a sea shore. If you stood on the shore of a lake and threw a rock into the water, the water would ripple. This is how sound moves in a ripple effect. Sound by definition is a movement of air through a solid, liquid, or gas. Imagine how these sound waves radiate 180 degrees off the speaker cone in your car.

Because sound waves bounce off reflective surfaces, the more places and spaces that they have to bounce off, the worse your sound will be in your car. If you can reduce the reflective surfaces, sound waves can better reach your ear in their pure form and the better the sound will be in your car.

You can solve these acoustical issues by minimizing the reflections around the speaker. Do this by packing the areas that are out of sight around the speaker vicinity with acoustical foam. To buy acoustical foam, I recommend

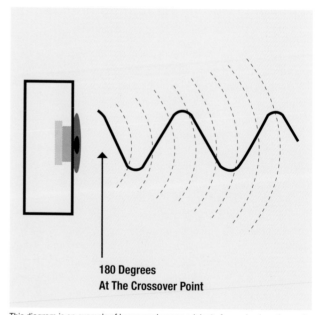

**180 Degrees
At The Crossover Point**

This diagram is an example of how sound waves originate from a loudspeaker and move through the air. The sound waves will radiate close to 180 degrees around the crossover point of the speaker. Crossover point means the particular frequency where the speaker starts playing music.

doing an online search of sites such as Amazon or eBay. A simple Google search should turn up some online retailers that can help you purchase this item. It is a little

Using an acoustic catch and/or a wood diffuser are more advanced methods of improving the interior acoustics of your car. This picture is of a wood diffuser. You would want to place a wood diffuser in a rear window that is not used for driving. Using this would stop the sound waves from reflecting off of the glass of the window.

Shown here is a piece of thin acoustical foam, which can be used to treat the area around where your speakers mount. Just cut the foam to the desired shape, apply spray adhesive, and press it into the area where you want it to sit.

costly, but well worth it. Use as thick of a foam as you can fit in the space. You can do this in unseen areas, such as under the floor mats or under the dash. You can also cover the speaker enclosure where your speaker mounts with cloth. Try to remove as much clutter (relays, wiring harnesses) from around your speaker as possible.

Finally, you can reduce reflections in your car by having more cloth surfaces. You can also build an acoustic catch (or absorber) or a wood diffuser. An acoustic catch is a thick piece of foam that is placed in a small, enclosed space. It absorbs the sound waves as they bounce around your car. You can place it in your rear window (if you have a show car that doesn't drive). The catch consists of foam and should be very thick (10–12 inches). A wood diffuser is the opposite of an acoustic catch and has sharp, wood slats that break the sound waves up so that they are inaudible and don't continue to bounce. Both of these techniques require a lot of calculation to determine the proper number of slats on your diffuser or the amount of foam for your catch, so make sure that you spend a lot of time planning for what will work best for your car and system.

It takes complex calculations to determine how you want to use these techniques. As you are learning about them, the general rule of thumb is that any foam that is 4 inches thick will absorb most of the early reflections

inside a car. So for your first acoustic catch, start by playing with 4-inch foam in various positions in the rear of your car windshield.

For more information on wood diffusers, try consulting a local recording engineer for help with designing your first wood diffuser for your specific vehicle. You'll want to do this simply because wood diffusers will differ greatly between makes and models of vehicles, as well as the type of system you are putting together.

SOUND PATH LENGTHS AND SPEAKER PLACEMENT

Inside a car, you sit off-axis from where your speakers mount. Most car systems have speakers mounted in the door or kick panel, and you sit in the car or passenger seat, which is 90 degrees to the speaker. Listening to sound broadcast in that arrangement is called off-axis response.

In the driver's seat, you will always hear the driver's side speaker louder than you'll hear the passenger's side speaker. This is because the distance from your left ear to the driver's door speaker is shorter than the distance from your right ear to the passenger side door speaker. These distances are called path lengths. The closer those two distances are to each other, the better your system will sound. This distance affects the amplitude balance and tonal balance of the system. As you move the speakers further in front of you (such as mounting them in the kick panels), the distances

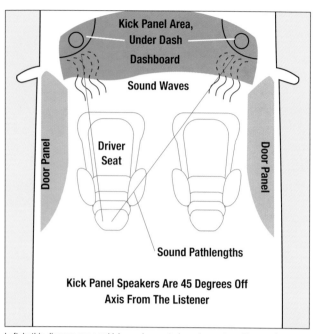

Kick Panel Area,
Under Dash

Dashboard

Sound Waves

Door Panel

Driver
Seat

Door Panel

Sound Pathlengths

**Kick Panel Speakers Are 45 Degrees Off
Axis From The Listener**

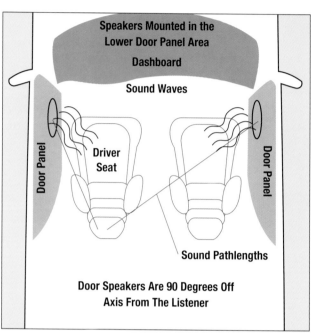

Speakers Mounted in the
Lower Door Panel Area

Dashboard

Sound Waves

Door Panel

Driver
Seat

Door Panel

Sound Pathlengths

**Door Speakers Are 90 Degrees Off
Axis From The Listener**

Left: In this diagram you see kick panel–mounted speakers, where the sound waves are originating 45 degrees off axis from the listener's ear. Notice that the path lengths shown in red are almost equal, which will provide you with a very balanced sound system and smoother frequency response. Right: In this diagram you see door panel–mounted speakers, where the sound waves are originating 90 degrees off axis from the listener's ears. Notice that the path length on the left side to the driver's seat is a lot shorter than the right side. This means that you will hear the sound from the left speaker in the driver seat before you hear the right speaker. This means you will have an unbalanced sound system.

This photo shows an acoustically treated interior with no reflections. Notice that there are no hard plastic surfaces, no shiny leather (which is reflective), and the entire interior is covered in acoustic cloth instead of factory materials. This replicates the factory look but keeps all of the surfaces nonreflective.

This is the interior of a Honda Civic Si that has cloth seats, carpeted floors, and a soft headliner with sound damping underneath. The painted door panels and other painted interior spaces add some reflections to keep the car from sounding too acoustically flat. *Joe Greeves*

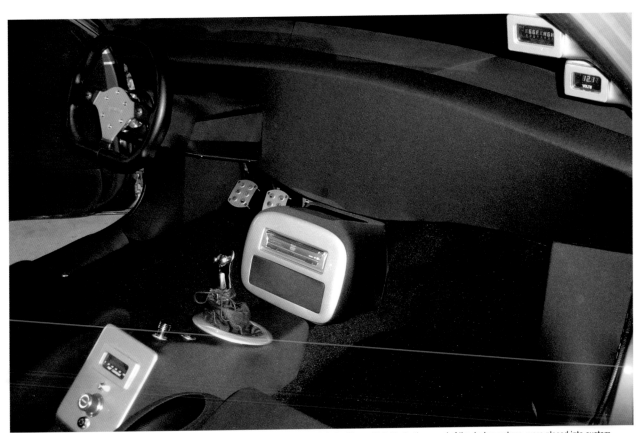

This acoustically treated interior has a reconstructed dash. The gauges from the dash and their wiring, which blocks the sound of the dash speakers, were placed into custom A-pillars so that the car can remain drivable.

between your ears and the speaker increase, but probably become more equal from side to side.

By mounting your speakers in the kick panel area, you're going to give yourself better sound balance and a better stage. If done properly, mounting the speakers in the kick panels can create a sound stage in front of the listener that extends in front of the car and even out onto the hood. This can create a sound stage out in front of you that makes the sound feel like it is playing to you from out in front of the car as if there was not a windshield there. Some systems in some cars can even produce lifelike instruments that layer front-to-back. This concept is called staging and imaging. Imaging describes vocals and instruments and the audible cues they produce that are almost visible. Staging refers to the realism of the sound stage your audio system can produce. The best staging scenario would be to have a stage with width and depth.

Ultimately, the best sound staging and imaging can be accomplished by positioning your speakers in kick panels with the speakers mounted as far away from the listener as possible. Where you want to exactly position your speakers varies on the vehicle's acoustics and the frequency response characteristics of the drivers that you choose. Try to get a speaker that has a wide dispersion pattern to mount into the kick panel area. Another cool tip is to get in the car with the speakers mounted to a

piece of wood, and move the wood around to different angles while you're listening in both seats. This will help you determine what the proper angle is for your car. You'll know they are in the right place when the majority of the vocals seem to be coming from the center of the dash when you're sitting in either seat.

One way to help keep excess noise out of your car is to properly create a chassis ground point for all of the electronics in the car. By doing this correctly, you reduce the chance of pops, clicks, alternator whine, or engine noise that is associated with ground loops from entering your system.

If you really want to get technical with controlling noise, you can lower the noise floor in your system by modifying the wiring inside of your electronic components, such as your amplifier. I don't recommend doing this without the help of an expert in this type of electronic engineering. However, if you are going to compete with your car for sound quality, I highly recommend taking on the challenge.

Some sound quality competitors spend their entire careers trying to determine the best speaker placement for their vehicles. Having your speakers mounted just one degree off-axis can cost you a competition. If you decide that you want to compete with your car for sound quality at a competition, try to find an experienced competitor to help you with positioning of your speakers and tuning of your system for your first time through.

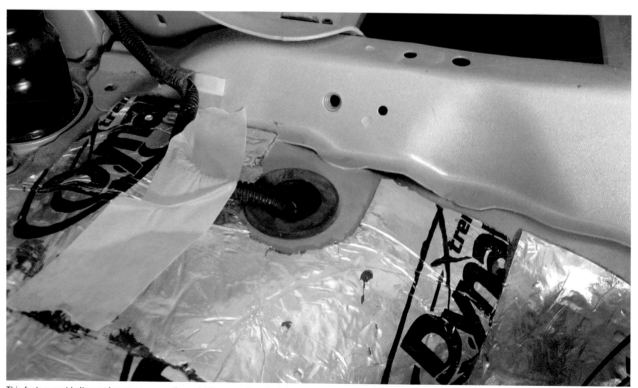

This factory seat bolt was chosen as a mounting point for grounding all of the system's components. This is a good point to pick because it is a solid ground point for the electricity to have somewhere to go.

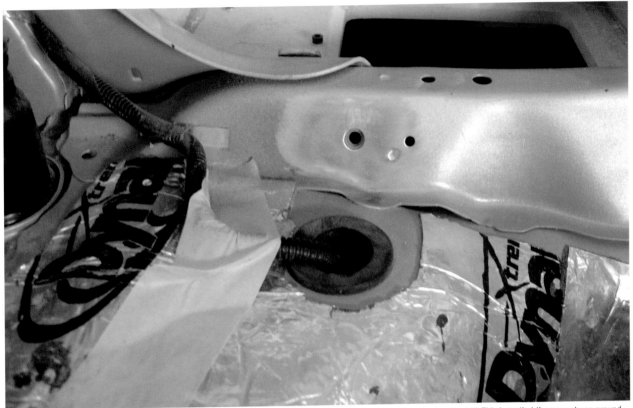

All of the paint on the mounting hole and a couple of inches around the hole should be removed using a wire wheel and a drill. This is so that the current can ground itself to bare metal.

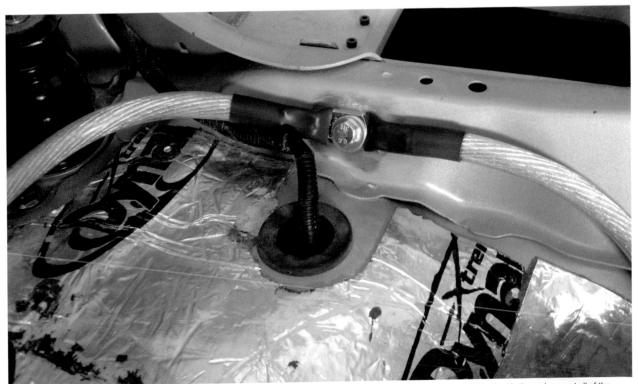

You should hook up large ground cables with copper ring terminals and bolt them to the ground point. Run these to your ground distribution blocks, and connect all of the electrical components of your stereo system to these blocks.

This photo shows the internal parts of an amplifier. Removing and replacing OP amps is the starting point to improve the sound quality and lower the noise floor in your sound system.

The next step in amp modification is to upgrade the power supplies in the amplifier and the capacitors, increasing the warmth of the sound and the detail in the music.

Chapter 5
Basic 12-Volt DC Wiring

A few years ago I was stuck in terrible rush hour traffic about six cars behind a red Mitsubishi Eclipse. This car was heavily modified and looked pretty cool. It had a great set of wheels, black as night tint, and a wicked spoiler. I doubt many people sitting in that jam missed seeing that car because what really drew attention to it was the boom coming out of it. Even with the windows up, you could hear that thing jamming from half a mile back. I always like to look at these cars and try to determine how good it sounds just from the point at which I can hear it.

As traffic picked up that day, the Eclipse accelerated off in front of me. I lost sight of it, but soon I could see smoke up ahead in traffic. The next thing I knew I was driving past the same Eclipse, which was now pulled over on the side of the road with a serious car fire. Three guys were running from the car right before it exploded.

Fortunately, in the case of the Eclipse no one was hurt. But stories like this one happen every single day. Now granted, they often don't have quite such horrific endings.

TIP!

As you begin to work on wiring in your car, you must be aware of certain safety precautions. Anytime a wire passes through metal, it must have a rubber-insulated grommet around the wire. I cannot stress this enough. If any type of wire passes through any type of metal—even through a drilled hole—you must use a grommet. If you don't, the wire can scrape and short—and you car can burn down completely. You would be shocked to know how many times I have seen cars burn down on the side of the road because of this safety error.

But many people who wire mobile electronics into cars often do so before they get their education in basic wiring.

Vehicles are full of electricity (voltage and current), flammable fluids, and many moving parts that create heat and friction. When you begin to modify the wiring within a car, try to keep in mind that every single wire, connection, and fuse can affect the car in many different ways.

The photos above show how to make custom grommets for a power wire to run through a firewall or any other kind of metal. At left are two pieces of black plastic that are screwed together with each piece on the opposite side of the firewall. The wire passes through the center. At center is a grommet made of metal with a compression fitting on the inside. The picture on the right shows the wire passing through a metal grommet, which is also in two parts, with either part on the opposite side of the firewall.

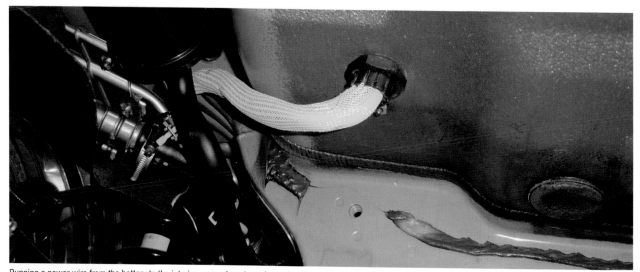

Running a power wire from the battery to the interior compartment requires going through the firewall. It is imperative that you use a rubber grommet to protect the wire. Preferably, use the point of entry where the factory ran its wiring from the engine to the interior. After you run the power cable through the grommet, be sure to add a protective sleeve, as shown here.

Now that you're ready to begin working on your system, you need to be sure that you have a good understanding of basic wiring. DC, or direct current, is the electrical current that flows through your car, using a battery as a conductor. Most modern cars function with a rechargeable 12-volt battery.

OEM ELECTRICAL SYSTEMS AND HOW THEY FUNCTION

A car's electrical system consists of the following major components:

1. Battery
2. Alternator
3. Starter
4. Fuse block
5. Relays
6. Solenoid

These components work as a system in the following way: First, the 12-volt battery stores its electric charge in a cell. This charge stores voltage and current that will be used to start your car. The starter draws upon this stored current to start the engine. The starter solenoid gets its power from the battery and receives an additional amount of current from the car's ignition on the interior. This action causes the starter solenoid to send a boost of current to the starter, which starts the engine.

Once your engine starts, it turns a pulley that turns a belt that is connected to the alternator. A second pulley on the alternator turns as well. The reaction of the two pulleys turning converts mechanical energy to alternating energy, or an electric charge for your battery. Thus the alternator continuously charges your battery. A typical alternator produces about 60 amps of charge.

All power cables should be covered in a loom or protective sleeve (conduit) to protect the wire. Additionally, all power wires coming off of a battery must be fused at the battery. All other 12-volt power cables (including accessory cables) must be fused at the source, except for capacitors, which I show you how to do later in this chapter.

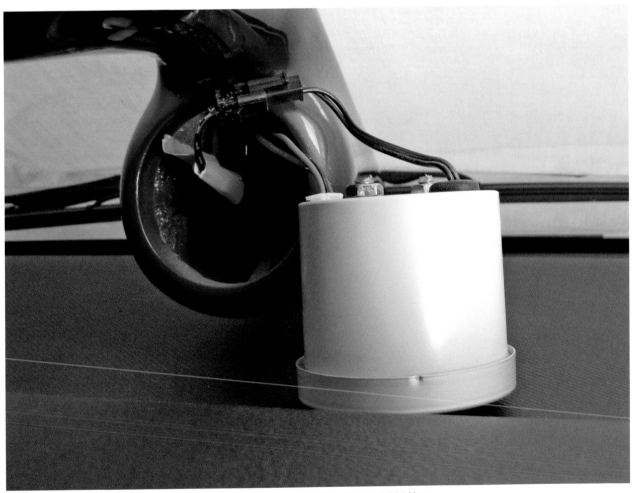

When wiring your aftermarket gauges, make sure you solder the connections and cover them with heat-shrink tubing.

Accessory Wiring

The rest of a car's electrical system runs off of the battery. All mobile electronic components and accessories wire to your car's electrical system, which gets its power from the battery.

Aftermarket accessories—including GPS, Bluetooth, satellite radio, and radar detectors—wire into your vehicle through the cigarette lighter plug in order to get power. Any kind of aftermarket gauges will hard-wire to the car's electrical system. Bluetooth, GPS, and satellite radio units need to not only be wired for power, but will need audio integration into your system as well. In Chapter 2 we discussed using an FM transmitter to connect the audio in your car to your mobile accessories. There are other options for connecting the audio in your car with mobile accessories that are better. For example, if your source unit has an auxiliary output setting,

then you can run your audio through that output. We will cover that option more in Chapter 6.

If you want to hard-wire your gauges to your car, you must first start by getting 12-volt power from your battery (called constant power) and then get ignition power from the fuse panel or steering column (called switched power). Because your gauges will normally be monitoring a system within the car, such as voltage, air flow, or temperature, you will need to look for the sending unit that comes with the gauge. This sending unit will connect to whatever part of your car's system that your are monitoring. For example, if the gauge is supposed to monitor your car's fuel pressure, then the sending unit will connect your fuel pressure regulator. Thus, after you connect power to the gauge, you would connect the sending unit to the fuel system and ground the wire.

Once you have all of your accessories connected and mounted, you must connect all of their power and ground wires to a power strip inside the car. A power strip, or barrier strip, allows you to run one large-gauge wire from a power supply to the power strip. The numerous setscrews on the strip allow you to connect all of your additional wires with ring terminals to share the power. Power strips can be purchased at Radio Shack or other electronics supply stores.

Here is a picture of a factory Molex plug (blue) connected to an aftermarket wiring harness (white) with the multicolored wires leading to the aftermarket source unit.

Source Unit Wiring

As I mentioned in Chapter 1, aftermarket source units use a wiring harness specific for your vehicle. You must buy the wiring harness for your car separately from your source unit. The source unit also has a wiring harness attached to its back side. When you remove your factory source unit from your car, you will be left with a single plug inside the dash. Your aftermarket head unit cannot connect to this plug, so you must use the separate wiring harness to adapt the plug so that the new source unit can be plugged in to the vehicle's wiring.

When you begin the process of connecting the new source unit's wiring harness to the wiring harness that matches your vehicle, you will need to pay close attention to the colors of the wires. In most cases, the aftermarket vehicle wiring harnesses are color coded in the same way as the wires on the aftermarket source unit. This is not always true; you can check the legend on the packaging for the vehicle wiring harness so that you can see if the wires' colors will correspond to your head unit's wires. In cases where the colors correspond, you will wire the harness to the source unit color for color: red-to-red, gray-to-gray, white-to-white, and so forth.

If your system is going to use amplifiers, you will use the preamp output on the back of your source unit, which is an RCA connection. Depending on the configuration of your

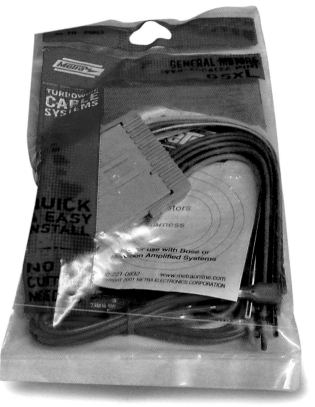

Most wiring harnesses are color coded, making them easier to connect.

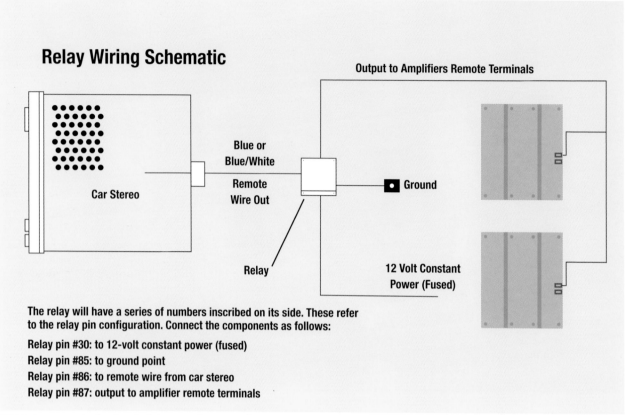

Relay Wiring Schematic

Output to Amplifiers Remote Terminals

Car Stereo

Blue or
Blue/White

Remote
Wire Out

Ground

Relay

12 Volt Constant
Power (Fused)

The relay will have a series of numbers inscribed on its side. These refer
to the relay pin configuration. Connect the components as follows:

Relay pin #30: to 12-volt constant power (fused)
Relay pin #85: to ground point
Relay pin #86: to remote wire from car stereo
Relay pin #87: output to amplifier remote terminals

Make sure you use a relay if you're going to have more than one amp in your car.

system and the locations of your speakers, you may use the front, rear, or sub outputs on the source unit. If you are going to place an amplifier on each set of speakers, then you will use all three outputs.

Relay Wiring

If you're going to wire a lot of aftermarket gear inside your car, you're going to need to use a relay. All amplifiers have a remote connection where they actually turn on or off. The source unit has a remote output that can be connected to the remote terminal on the amplifier with 16-gauge wire. This output on the source unit is suitable for turning one amplifier on or off. If you're using more than one amplifier, then you'll need to use a relay.

A relay is a simple device (usually about $5 at most electronic supply stores) that takes the low-amp remote output from the source unit and uses it as a trigger to send 30 amps of output to turn on all of your extra devices. The low-amp output of the source unit is not enough to turn on two or more amplifiers by itself. The relay will take the low-amp current and convert it to high-amp current.

Capacitor Wiring

In Chapter 1 we discussed what a capacitor is and how it can be used with a car audio system. Recall that a capacitor needs

to be mounted as close to your amplifier as possible. Finding a place to mount your capacitor is often far more challenging than wiring it. You only use a capacitor if you are using an amplifier, and wiring an amplifier requires running a main power cable directly off of the battery.

To wire your capacitor, take your main power cable from the battery and connect it the positive terminal on the capacitor first. Then run the same size power cable from the capacitor to the amplifier's positive terminal. The capacitor does not need to be fused. The ground terminal on the capacitor is connected to your amplifier's ground point with the same size cable.

After your capacitor and amplifier are connected, you must be sure to charge them before you use your system. The **main** power cable coming off the battery that is supplying the power to the capacitor and amplifier needs to be fused as close to the battery as possible. (Look a little further in this chapter to see how to fuse a wire.)

Before you charge your capacitor, it is important that I stress to you the danger of doing it improperly. If it is done wrong, you risk serious harm! I strongly suggest using a resistor across the fuse terminals at the fuse point under the hood.

By using a resistor, you create a trickle of power to the capacitor instead of a full force of power. If the capacitor gets

This is a typical Bosch relay, and it's good for 30 amps of output. Make sure the power and ground wires for the relay are capable of 30 amps.

too much power too quickly, it causes a huge, explosive spark that can be very dangerous. Once you have installed your capacitor and connected all of the wires, and then have used a resistor to charge it, you can then insert the fuse into the fuse holder safely and remove the resistor. You would then remove the resistor first and then replace the fuse, but only after you have a full 12 volts at the capacitor. Make sure that you use a meter to measure the volts!

Amplifier Wiring

As we just stated, you have to run a power wire from your battery to your amplifiers and then carefully fuse this wire under the hood. In addition to this main power wire, you will need a main ground wire of the same gauge size. The ground wire connects to a chassis ground point in the vehicle. I prefer to use a seatbelt bolt or seat bolt as the chassis ground point. The concept is that you remove the bolt and grind the paint under the bolt (with a wire wheel or brush) until it is bare

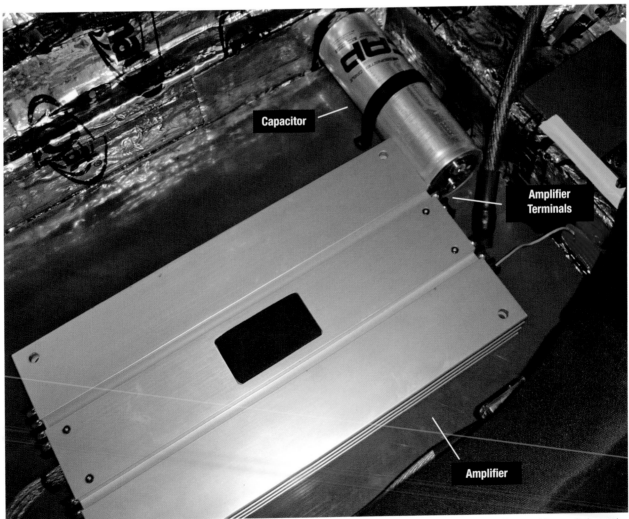

As we have already discussed, capacitors need to be mounted as close to the amplifier as possible. Capacitors do not need to be fused. You do need to connect the capacitor's ground wire to the ground point in the car where the amplifier is grounded. This photo illustrates a capacitor mounted directly next to the amplifier's power terminals.

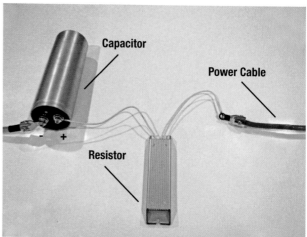

Capacitor

Power Cable

Resistor

- +

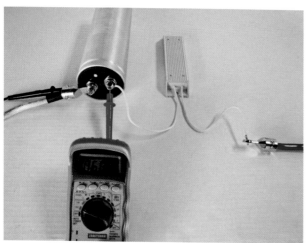

These pictures show the steps for correctly charging a capacitor. First (starting at the top left) you identify the positive terminal. Then (2) run the power cable that will power the capacitor (and comes from the battery) into a resistor. The output side of the resistor connects to the capacitor's positive terminal. Next (3), connect the ground to the negative terminal. Then (4) use a multimeter set to VDC to measure the voltage across the terminals of the capacitor. Once the multimeter reads 11 volts or higher, you can remove the resistor and connect the power cable. Shown here at the bottom are the final connections on the capacitor. Never skip to this step and connect the power directly to the capacitor, or it can cause the capacitor to spark.

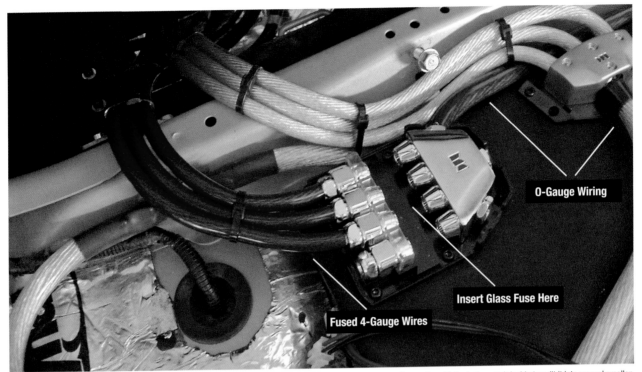

O-Gauge Wiring

Insert Glass Fuse Here

Fused 4-Gauge Wires

If you use a distribution block, you must observe an important safety rule. Many people will take a large power wire coming off of a battery and decide to split it into several smaller wires to distribute the power throughout their components. Most people think to fuse this one larger wire with a 100-amp fuse. However, if you are using a distribution block to split a cable to smaller sizes, you must use a fuse that the smaller wire will be able to trip. This is because the small wire is connected in parallel to the battery. If it touches a ground and you have a 100-amp fuse in there, you won't pop the fuse. So be sure to use the appropriate size fuse for the smallest wire that you have coming off of the block. A power distribution block simply splits a cable into multiple cables. A fused power distribution block does the same thing but also adds a fuse to protect the split wires.

metal. Connect your ground wire with a ring terminal to this bolt and retighten the bolt in the car. The other end of the wire runs to your amplifier's negative terminal.

If you're using more than one amplifier, you need to split the power and ground wires from one cable to the number of amplifiers you are using. So if you have three amplifiers, you will split the cables from one to three. You can do this by using a distribution block.

If you use a distribution block, you must observe an important safety rule. Many people will take a large power wire coming off of a battery and decide to split it into several smaller wires to distribute the power throughout their components. Most people think to fuse this one larger wire with a 100-amp fuse. However, if you are using a distribution block to split a cable to smaller sizes, you must use a fuse that the smaller wire will be able to trip. This is because the small wire is connected in parallel to the battery. If it touches a ground and you have a 100-amp fuse in there, you won't pop the fuse. So be sure to use the appropriate size fuse for the smallest wire that you have coming off of the block. A power distribution block simply splits a cable into multiple cables. A fused power distribution block does the same thing but also adds a fuse to protect the split wires.

Shown are two Monster Cable power distribution blocks. The one on the left splits one O-gauge to four 8-gauge wires. The one on the right splits one O-gauge to three 4-gauge wires.

The left and right RCA cables (above) get connected to the left and right audio input on the amplifier.

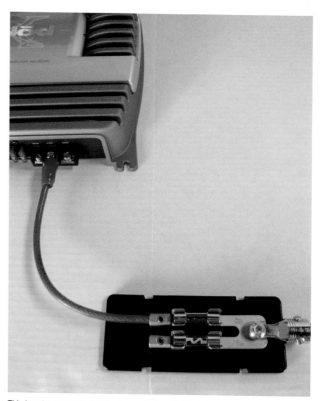

This is a 4-gauge wire coming from a battery that is being split to go to two amplifiers. It comes into the fuse block as a 4-gauge wire and comes out as a fused 8-gauge wire that goes to the amp.

The RCA cables (below) plug into the RCA output of a source unit.

This is a standard pair of Monster Cable RCAs with left and right connectors.

The RCA connections that we previously discussed on the back of your source unit connect to your amplifier's RCA input connections. At the source unit, you'll have a left (white) and a right (red) connector. You will find the same left and right (white and red) connection at the amp. You can connect the two of these with an RCA cable.

Your speakers connect to your amplifier's positive and negative output terminals. The left positive and negative connect to the left speaker's positive and negative terminals; the right positive and negative connect to the right speaker's positive and negative terminals.

Typical Single Voice-Coil Subwoofer Wiring

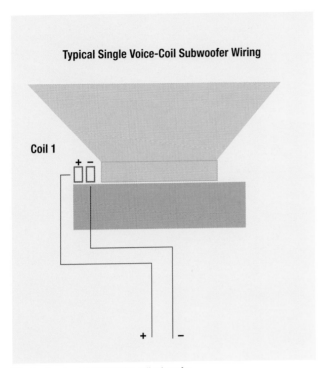

Coil 1

+ −

+ −

This drawing shows a single voice-coil subwoofer.

Subwoofer Wiring

Subwoofers are available in single and double voice coils and a variety of impedance ratings. Most subs connect to your amp in mono or stereo. Stereo is a straight connection of positive and negative for the left channel and a positive and negative connection for the right channel to an amplifier, just as a coaxial or component speaker would connect to an amplifier.

A stereo connection to an amplifier represents a left and right connection. A mono connection is when you take the two stereo connections and bridge them together to make one mono output channel. A mono connection is what you would use if you need to get more power out of the amplifier. To do this, you must make sure that your amp is bridgeable, and be sure to find out how low of an impedance your amp can go and still remain stable. Most car audio amplifiers are stable at 2 ohms, which means this is the lowest impedance that amp can see and still perform well. Connecting single voice-coil subwoofers to an amplifier is quite simple to do. Once you have bridged the right and left channels on the amplifier to a single channel, you can then connect the subwoofer to this channel with positive and negative. If you have an 8-ohm sub, it will give you an impedance of 8 ohms at the amp. If you have a 2-ohm sub, it will be an impedance of 2 ohms at the amp.

When you're purchasing an amplifier, there will be power ratings for it on the box. For example, a common rating is 200 watts at 4 ohms. Two single voice-coil subwoofers can be wired in parallel or series connection to an amp, or a dual voice-coil subwoofer can be wired in parallel or series connection to an amp. Parallel means taking the positives from the two subs or from the dual voice-coil sub and connecting them together, and then doing the same with the negatives. You then connect these wires with the positives and the negatives on the amplifier. A series connection means that you connect the negative from one of the single-coil subs or one coil from the dual-coil sub to the positive of the other single-coil sub or coil from the dual sub. Then you take the remaining positive and negative and connect them to the amp.

A series connection of two 6-ohm subwoofers would give you 12 ohms at the amp. A parallel connection of two 6-ohm subwoofers would give you 3 ohms at the amp.

You would want to use the parallel connection to lower your impedance and gain more power out of the amplifier. You want to use a series connection when you want to raise your impedance, such as if you had purchased subwoofers that have a native impedance that is too low for the amplifier when they're paralleled together. All amplifiers have a certain impedance point that they are stable at. If you go below this impedance of the amplifier, it can damage the amplifier; so you would use the series connection to raise the subwoofer's impedance back to where the amplifier has a stable load present.

Video Monitor Wiring

Once you've got your sound wired, you're ready to start wiring for video. Video monitors consist of a power supply box that wires just like an amplifier (power, ground and remote) and a connection cable that goes from the power supply box directly to the video screen. The power supply box can be mounted away from the video monitor and hard-wired directly to the car. The connection cable then runs directly to the screen. The screen will receive power when you turn on your radio, as it is a remote output.

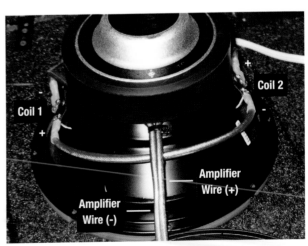

Coil 1

Coil 2

Amplifier Wire (+)

Amplifier Wire (−)

This subwoofer has two wires coming in from the amplifier, which is split and connected to each coil on the subwoofer. This is an example of a series connection. The remaining positive and negative terminal on each coil are connected together.

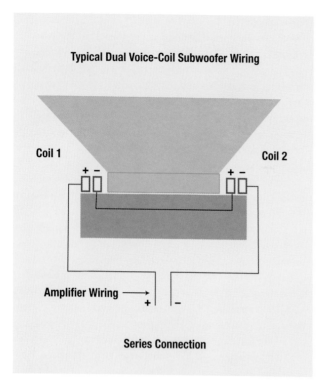

Typical Dual Voice-Coil Subwoofer Wiring

Coil 1   Coil 2

Amplifier Wiring ⟶

Series Connection

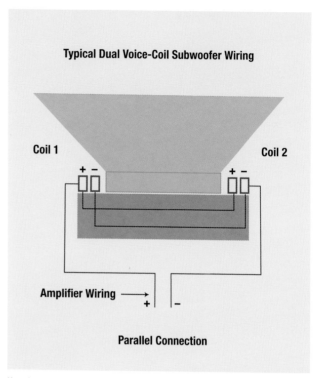

Typical Dual Voice-Coil Subwoofer Wiring

Coil 1   Coil 2

Amplifier Wiring ⟶

Parallel Connection

Here is another look at how to wire a series connection. A mono amplifier has one positive and one negative terminal for the wires to connect to. However, a dual voice-coil subwoofer has two positive and negative terminals, one set on each coil. The single positive wire coming from the amp connects to a single positive terminal on one side's coil. A single negative wire runs from the amp and connects to the opposite coil's negative terminal. The remaining positive and negative terminals on each coil are then connected with another wire to complete the series connection.

Here is an example of a parallel connection. This means the voice coils of a dual coil speaker are paralleled together. To do this, you would run one wire from the positive terminal on an amp to one positive terminal on one coil. You would then run a second wire from that same positive terminal on the coil to the second coil's positive terminal. You would repeat this process for connecting the negative terminals, creating a parallel connection between the positives and negatives.

Shown here are the power supplies for video monitors, which should be wired first with the video input coming from your DVD player.

Wiring Additional Batteries

Wiring additional batteries to your system can be a little tricky, but you do have a few different options.

If you recall, in Chapter 1 we discussed putting a starter battery under the hood and a deep-cycle battery in the trunk. This option gives you reserve current storage close to your amplifiers. You can isolate the rear battery from the front battery so that if you drain it while playing your stereo system at a car show or at the beach, the battery under the hood is still charged to start the car.

You can isolate a battery with a battery isolator. This device isolates the front battery from the back battery until you turn your key on. For example, let's say you have your stereo hard-wired to turn on without your car being on, and you're playing your system at a show. You can drain the rear battery until you turn your key on to leave. Once you turn the key on, it will activate the isolator and connect the batteries back together so that the alternator can charge your trunk battery. You can buy a battery isolator from Northern Tool for around $40. Be sure to buy a high-amp isolator that will accommodate your stereo system.

The cable connects the video monitor to its power supply before it snaps into its housing.

Another way you could incorporate dual batteries is to use a marine battery selector switch. A company that makes these is called Perko, and they're available for around $40. This heavy-duty switch has a dial that flips between battery one, battery two, or both batteries. This type of switch can be mounted anywhere inside the vehicle. Be sure to follow the wiring instructions included with it. You always want to follow instructions for wiring when they are available, if for no other reason than simply to double-check your steps. Remember, if something isn't wired correctly, you can risk burning down your car—and no one ever needs to have that happen to them!

Sometimes the factory wiring on a battery is too small, and your headlights will dim when your bass hits. You can fix this by upgrading the wiring on your battery with better connections.

There are different ways that you can charge your batteries. The most common way is to use a power charger, but then you have to connect these big alligator clamps on your battery terminals, and then you can't close your hood. A more practical solution is to use a power supply and install forklift quick-release connectors on it and your car's wiring that goes to your battery.

Marine battery selector switches can help you control multiple batteries for your car's system.

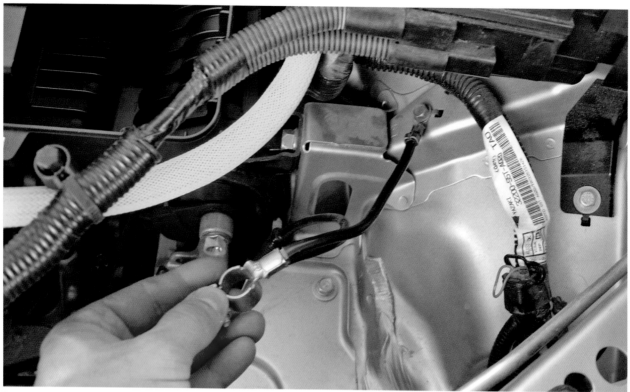

This is a factory battery ground wire that needs to be upgraded to as large of a wire as can be fit on the connector. This will keep the dash and headlights from dimming when the system is playing.

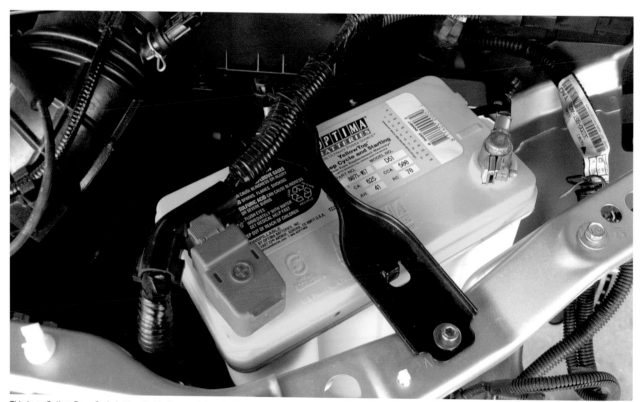

This is an Optima Deep Cycle battery that is the exact size of a factory import battery, usually specific to Honda and Toyota. Attached to this upgraded battery is upgraded ground wiring to help the car's lights function better with the system. It is always ideal to upgrade both the battery and the ground wire.

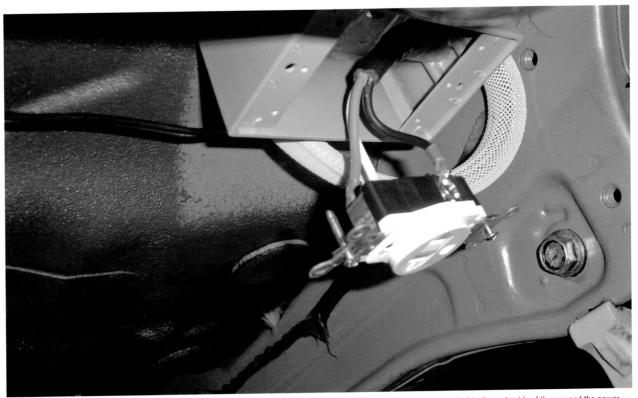

You can make a permanent power supply power plug connection underneath your car. This is a weatherproof electrical box bolted to the underside of the car, and the power cord from the power supply is hardwired to an electrical receptacle.

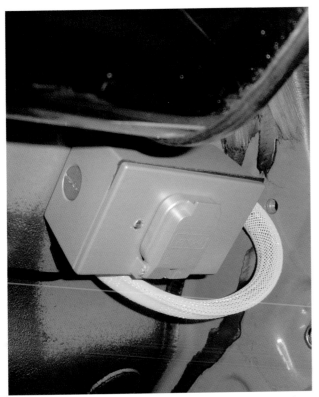

The installed receptacle needs to have a weatherproof cover to protect it from the elements.

The power wire should be run through a protective sleeve and plastic grommet, and the wire passes into the interior of the car to turn on the power supply.

This power supply, made by Cascade Audio Engineering, is the kind I like to use in my own cars. This model number is APS-90, and it provides a rock solid 90 amps of current and is regulated, so it won't burn up your batteries.

This picture shows the forklift quick disconnect. These are available at most forklift service centers, and you can get a pair for about $30.

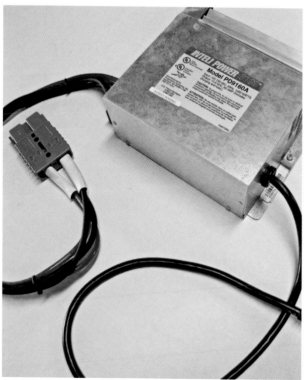

This is a less expensive power supply option that comes in a small, portable case. It is an Inteli Power 9100, which can be bought at Metra for about $175. You place the forklift connector on the output wires from the power supply.

The other forklift connector should be mounted to your car in an accessible spot. Then you can connect your battery cable to it. This way whenever you need to charge the car, you can plug in the two connectors and then the power supply.

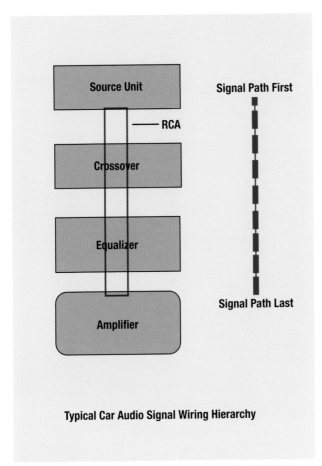

Source Unit

— RCA

Crossover

Equalizer

Amplifier

Signal Path First

Signal Path Last

Typical Car Audio Signal Wiring Hierarchy

When wiring both a crossover and an EQ, make sure you run the signal from the source unit to the crossover first and then the RCA output to the EQ.

Here is an example of a power wire (the yellow wire) in a protective sleeve run under the chassis of a car and connected to the frame in a secure spot.

BASIC 12-VOLT DC WIRING

EQ and Crossover Wiring

EQs and crossovers both have a power wire, ground wire, and remote wire and are wired in the signal path between the head unit and the amplifiers. You want to wire the signal from the source unit into the crossover first and then run the output RCA to the EQ input RCA. After that, you run the EQ output RCA to the amplifier. This scenario works when you are using both an EQ and a crossover.

If you're just using the crossover, you would wire from the crossover directly to the amplifier. Likewise, if you were just using the EQ, you would wire from the EQ directly to the amplifiers.

Power Wires and Fuses

Wiring multiple amps and additional electronics inside your car can seem overwhelming at times, but if you have a good plan, you will be able to get it all done with success. The main power wire that you run throughout your vehicle is the lifeline of your system. This power wire has to be of the appropriate gauge, based on the current requirement of your system. So you must take all of your amps and add up how much current they're going to draw from the electrical system. You will then have to make certain that your power wire will be able to supply that amount of current.

To find the appropriate wire size for your system's power cable, go online and search for AWG tables. In those tables you can find current carrying totals, which will cross-reference a wire gauge size that will supply the total current for your system.

In IASCA's official competition manual (*IASCA*, 1999), the formula for determining the size of fusing and power cable that you need is: Total RMS output of the amplifier multiplied by 2, divided by 13 (which is the average power available inside a car), and that equals the current draw in amps. For example, if your amp has 1,000 watts of power, times 2, divided by 13V, it would equal 152 amps. So you would then need at minimum a 175-amp fuse. When you cross-reference this on an AWG table, you will find that a 175-amp fuse equals a 0-gauge wire.

Always use the available online tables and the IASCA book to help you determine what size fuse and power cables you will need to use in your system. Please note that most amplifiers have internal fuses as well to help provide extra protection.

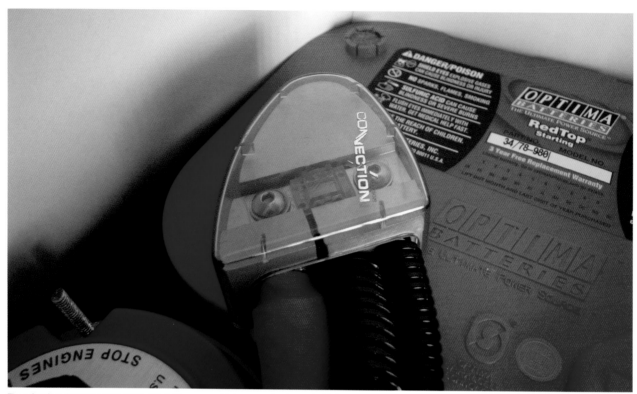

The safest fuse that you could possibly use for the power cable in your system is one that is integrated into the actual battery terminal itself, connected to the battery—as shown here in this photo. This photo shows a battery terminal that you connect your power wires and factory vehicle wires to. Notice the fuse in the center of the battery terminal; this protects the main power wire from shorting the electrical charge off of the battery.

This is another, more traditional way that you can protect the power wire that runs to your amplifiers. This shows an install of a circuit breaker in-line on the main power cable. It must be located within 18 inches of the battery and must be in-line on the wire before the wire passes through any metal opening.

Chapter 6
Mobile Electronic Accessory Installation

MOUNTING A SATELLITE RADIO

In Chapter 5 we discussed wiring your mobile electronic accessories in your car by utilizing the cigarette lighter connection or by hard-wiring them to your car. Once you have made a decision about how you're going to wire your accessory, you will need to decide where you're going to mount the device in your car. Let's walk through an installation of a Sirius satellite radio into a 2004 Honda Civic.

Step 1: The first thing you need to do when you get a new accessory is unpack the box and check the parts list to ensure that all of your pieces are present and ready to go into the car.

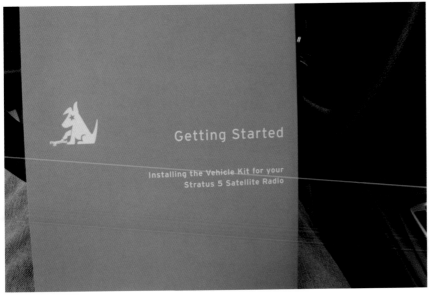

Step 2: Most electronic devices and accessories come with a "getting started" pamphlet. It's a good practice to read this document to familiarize yourself with the setup procedure.

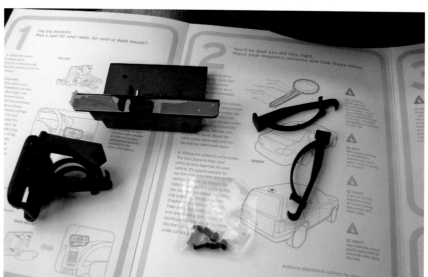

Step 3: This is a Sirius satellite radio display's mounting dock and bracket. The satellite radio plug-and-play kits usually have two mounting options. You can either use a dash suction cup or adhesive mount, or you can use an AC vent clip mount. In this case, I used the clip mount.

Step 4: Four set-screws attach the mounting bracket to the radio dock. An all-in-one tool (such as a Leatherman Wave) is useful for the assembly of mobile electronic devices. A Leatherman is a quick-access tool that gives you basic tools at your fingertips at all times. Of course, you would use it as a complement to a larger collection of tools, but a Leatherman is something you should always keep in your pocket.

Step 5: Once you have assembled the mounting bracket and the dock together, you need to test-fit it on the AC vent to make sure it will fit and stay stable when you're driving. You also need to check and be sure that you will still have access to all of your steering controls (such as your blinkers) and that your steering wheel or any other mechanism of the car will not be prohibited from functioning properly because of the device.

Step 6: All Sirius satellite radio kits come with an externally mounted antenna. This antenna must be placed outside the vehicle away from the other antennas. It needs to have a minimum of 3 inches of metal around it in all directions. The antenna wire must be routed to the interior of the car and should be placed underneath the weather-strip material around the front or rear windshield. You have the option of mounting this in the front or rear of the car.

Step 7: I placed the antenna in the center of the roof, as shown in this picture. Once it is placed where you want it, you can begin running the wire.

Step 8: A flat, plastic putty knife should be used to gently lift the factory weatherstripping so you can tuck the antenna wire underneath.

Step 9: Now the cable is under the weather-strip material down to the bottom of the windshield and into the engine compartment in preparation for running it inside the car.

Step 10: At this point, you must search the firewall of the car and look for a thick rubber grommet (usually circular) where wires run through the center of it. This will be your point of entry into the interior of the car.

Step 11: A long zip tie (around 3 feet) is a useful tool. Push the zip tie through the rubber grommet. Then go around to the interior of the car and look under the dash to see if you can see the zip tie pushed through. If so, then you know you have a clear entry into the interior. If not, then you need to find another grommet and try again. If there's a large bundle of wire going through a grommet into your interior, then 99 percent of the time it will be your best point of entry.

Step 12: After confirming that your zip tie pushed to the interior, use electrical tape to attach the antenna wire to the zip tie, and then stick it through the hole in the grommet.

Step 13: Remove the plastic panel underneath the dash to access the zip tie and the antenna wire and then pull it into the interior of the car.

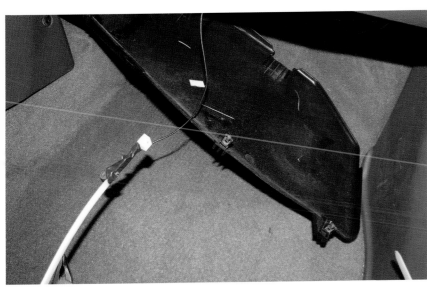

Step 14: Make sure that you pull the zip tie and antenna cable gently and completely into the interior of the car. Also ensure that you connect the antenna wire and the power plug to the device's dock. Use zip ties to help tie up and hide the antenna wire under the dash so the wire is safely out of the way.

Step 15: The power requirement for this electronic device is a standard cigarette plug, as shown in this picture.

Step 16: Connect the radio face to the dock when all of the wires have been connected.

Step 17: As a secondary option, you can mount the radio to the center console, making it ergonomically easier to reach it from both the driver and passenger seats.

MOUNTING A LAPTOP

As we touched on in Chapter 2, installing a PC or laptop in your car is another way to integrate a mobile accessory.

Allow yourself about three hours total installation time when installing your mobile electronic accessories. A lot of people are intimidated at the thought of installing these sorts of devices, but once you get the first one under your belt, you will see that it is really a simple installation. Once you're done, you will be rewarded with a fun addition to your already awesome sound system.

Let's take a look at how to set up an external laptop with Internet access in a car.

Step 1: The first thing to consider when you're trying to integrate a laptop or any external computer in your car is where and how it will be mounted. The absolute best mounting system for a laptop is made by Ram. They are made for police vehicles and can be purchased at www.ram-mount.com. In this picture, I used a generic-brand police-issue laptop stand, and on this bench you can see the build quality of the stand and all of the parts that come with it.

Step 2: Make sure you choose a laptop stand that has a multipoint adjustment, such as the one in this photo. This will enable you to adjust the position of the laptop in the car, helping to control the glare from various types of sunlight.

Step 3: The laptop stand must be securely bolted to the floor of the vehicle with 3/8-inch bolts.

Step 4: Once you have the laptop stand securely bolted to the floor of the vehicle, position the stand where you can still ergonomically reach the stereo and AC controls in the dash. You also want to make sure you don't obstruct your view through the front windshield when the laptop is open.

Step 5: Use a 12-volt DC converter to convert the car's voltage to 110 volts so that you can plug the laptop in and get power.

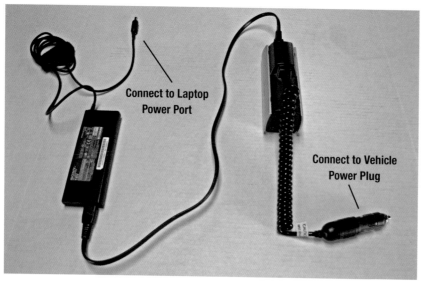

Connect to Laptop Power Port

Connect to Vehicle Power Plug

Step 6: Here is an illustration that shows how to connect a standard laptop AC power plug to a DC to AC power converter.

Step 7: Use a USB Internet card from your cellular provider, and install it on your laptop. This will give you full Internet access while you're on the go. Most cellular companies have monthly plans for this type of service.

Step 8: The wireless USB card gets installed on the laptop. At all four corners of the stand, there are flip-up tabs to securely hold the laptop in place when the car is moving.

Step 9: Here you can see the final set up of the laptop in the car. Remember to follow all recommended safety precautions when using this sort of technology while operating a vehicle.

LIGHTING

Lighting is a great way to draw attention to your car, especially if you add it to the exterior of your car. LED lights are very bright, so if you have them on while you drive at night you can be seen as a bright blur from very far distances. Let's take a look at how to add this type of lighting to a car.

Step 1: This is a 48-inch LED tube that is being placed underneath a car on the edge of the frame rail. This placement protects the tube, in case something flies under the car. If that happens, the rail gets hit instead of the tube. You will need to wire the lights to the neon controller that connects to the battery (see Chapter 2). Each lighting kit wires differently, so be sure to check the instructions that come with the kit to make sure you do it correctly. Usually, there is one wire that goes into the tube and one that comes out from the tube, and all wires are in series together. In this scenario, the power would come from the car's battery into the controller, out from the controller to the first tube, and then over to the second tube.

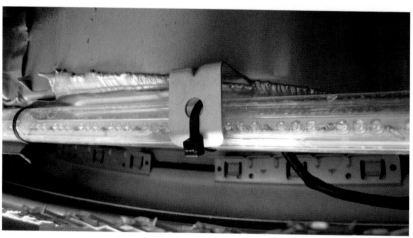

Step 2: Whenever possible, use factory mounting tabs or brackets to secure the LED tube to the car. This will keep the tube from moving around while you drive.

Step 3: Once you have secured the tubes to your vehicle, power the tubes and rotate them until you have the desired lighting effect.

Step 4: This is a completed LED tube undercarriage installation with the power on, to demonstrate the glow.

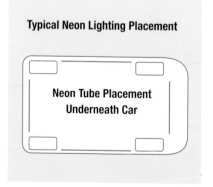

Typical Neon Lighting Placement

Neon Tube Placement Underneath Car

Step 5: This chart shows the most optimal placement point for neon/LED tubes underneath a car. I always place the shorter tubes horizontally across the car left to right, and I run the longer tubes vertically front to back. I also try to mount them far enough under the car so that you can't see the tube; you can only see the glow from the light.

Chapter 7
Speaker Installation

Believe it or not, the placement of the speakers in your car's audio system can make a huge impact on how it will perform. If mounted poorly, your speakers will just not sound that great. Mounted properly, your speakers can produce sound that simulates a real-life band. In this chapter, we're going to explore some of the different aspects of speaker installation, and we'll examine a few different ways to go about mounting speakers and subwoofers.

UPGRADING DOOR SPEAKERS
When installing aftermarket speakers in your doors, often you will find that they won't drop right in. This is because most aftermarket speakers are deeper than your factory speakers because of their oversized magnets.

You may have to make a spacer ring to adapt the new speaker to the factory location. Do not make a spacer ring out of wood. Doors are very prone to moisture, and moisture causes wood to swell. Always make your ring out of plastic or metal. If for some reason you don't have the ability to do metal or plastic, coat a wood ring with a mixture of fiberglass resin and allow it to dry. This will help to seal the wood and keep it from swelling so much.

No matter what type of material you use, I suggest spray painting the rings black. This not only makes them look better behind the door panel, but it also creates an additional barrier to moisture.

To demonstrate changing the factory speakers in a car to aftermarket ones, let's take a look at a Mercury Cougar I worked on a few years ago.

Step 1: The first step when changing door speakers is to gently remove the factory door panel. When you do this, be careful not to snap off the clasps that connect the panel to the car door. Use a flat putty knife to help you remove the door panel by inserting it between the panel and the door plate on the outer edge of the panel.

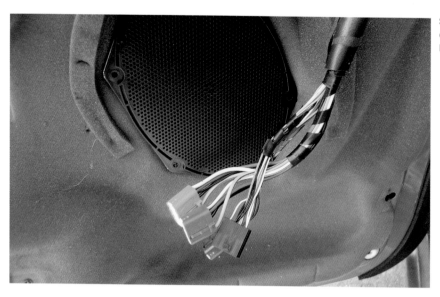

Step 2: Remove the OEM speaker from the car door—there are usually four bolts or screws that hold it in place.

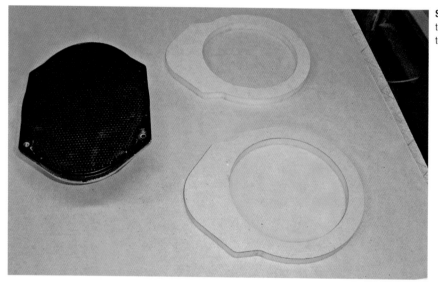

Step 3: The OEM speaker can be used to create a template for your speaker rings. Templates are easiest to make out of MDF wood.

Step 4: Create wood templates for your speaker rings for both car doors using 1/2-inch MDF. The factory speaker had a 1/2-inch lip on it, so you must use wood that is the same thickness as that lip for your template.

Step 5: Brush fiberglass resin onto the wood speaker rings to seal them against moisture, and add a heavy coat of black spray paint to help seal the resin.

Step 6: Install the wood ring onto the door panel using the factory screws. Then mount the 6.5-inch aftermarket speaker onto the wooden ring.

Step 7: Mount the 1-inch dome tweeter component to the factory sail panel with the provided 6x32 bolt and wing nut that comes with the tweeter. This completes your speaker installation.

As you can see, changing out your factory door speakers only takes a few steps and can usually be done in a few hours for about $10 in materials (plus the cost of the equipment).

KICK PANEL INSTALLATION

Kick panel enclosures provide the best sound quality and performance from your car's aftermarket speakers. The reason this is true is because they can help you better equalize the path lengths from your ears to the speakers. If you can get the distance from your left ear to the left front speaker and the distance from your right ear to your right front speaker within 7 to 10 inches of being the same, you will have a magnificent sound stage.

Adding kick panels to your audio system is a must if you plan on doing any professional competitions with your system. They can often be one of the most technical pieces of your system, so you need to make sure that your build is precise and clean. It may take a few tries before you can perfect building kick panel enclosures, so don't get discouraged. Just keep practicing, and you will get some great sounding enclosures.

If you decide to build kick panel enclosures for your speakers, it's important to make sure that they don't become a driving obstacle. You also need to pay attention to where your hood latch is located so that you don't build your kick panels over your access to it. Try to make sure that you will be able to place the speakers at identical angles without impairing the car's drivability or functionality.

Let's take a look at some kick panel enclosures being built for a 1998 Chevrolet Silverado.

Step 1: The front floorboard of a truck is where I installed 6.5-inch component set speakers in kick panels.

Step 2: A small amount of sheet metal was cut out of the kick panel area with an air saw to allow the speaker magnet to sit flush.

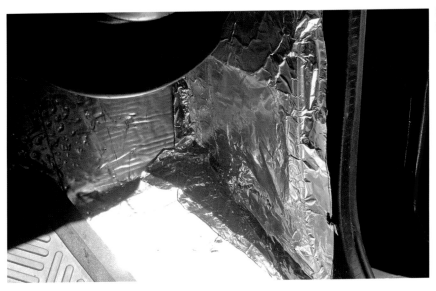

Step 3: Aluminum foil was spray glued into the kick panel area and smoothed out to create a template.

Step 4: The aluminum foil was marked out with a Sharpie and trimmed to the desired shape.

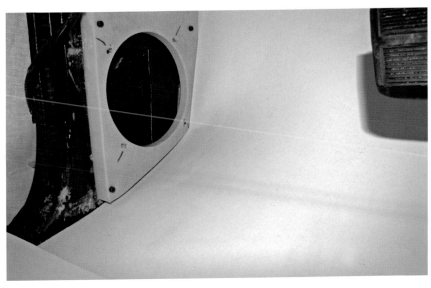

Step 5: The aluminum foil was transferred to 1-inch plastic, and the kick panel shape was cut of the template with a jigsaw and then routed on a router. Here, you can see the cut plastic shape mounted to the truck.

Step 6: I test-fit a 6.5-inch speaker in the mounting baffle.

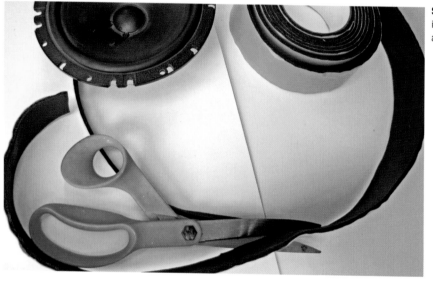

Step 7: Adhesive-backed weather-strip tape was cut into a 1/2-inch-wide strip to form a weather-strip seal around the mounting flange for the speaker.

Step 8: The weather-strip material was applied to the back of the 6.5-inch speaker.

Step 9: Black carpet was glued and wrapped around the mounting baffle.

Step 10: The center section of the carpet was removed so that the speaker could mount to the baffle, and white pillow stuffing (or Dacron) was stuffed in the opening to sound damp the speaker.

Step 11: Shown here is an assortment of heat-shrink tubing, which is sold as a kit from Harbor Freight Tools. This is essential to add to all speaker wire connections because it insulates the wire and connector so that they can't come apart.

Step 12: A cordless torch on low power was used to melt the heat shrink.

Step 13: The speaker wires were zip tied together to prevent them from rattling inside the kick panel enclosure.

Step 14: Be sure to crimp connectors on your wires securely and then gently squeeze the female connector opening before sliding it onto the male speaker terminal.

Step 15: The mid-bass component speaker was bolted to the kick panel with four 8/32-inch bolts.

Step 16: The Sony tweeter from the component set was mounted, and a polypropylene capacitor was soldered to the tweeter to protect it.

Step 17: This is another sealed kick panel from a different vehicle that housed a Dynaudio MW150 midrange and a Dynaudio MD130 3-inch tweeter.

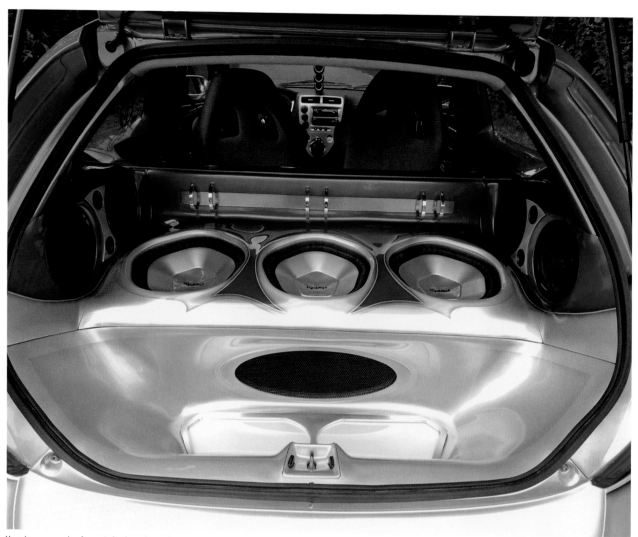

Here is an example of a ported subwoofer enclosure. *Joe Greeves*

DESIGNING A SUBWOOFER ENCLOSURE

Subwoofer enclosures can be sealed, vented, band-pass, or free-air:

Sealed: The woofer is mounted flush to a sealed enclosure, with the magnet and the back of the cone facing into the box. When the speaker plays music, it moves, and the air inside the box compresses and it pushes the cone back out toward the front of the car.

Advantages: Bass is tighter, and they can be built small. There's a 15 percent margin of error, which means that you can be up to 15 percent off in your measurements and still have the sub perform well. They can handle more power and have a smoother frequency response.

Disadvantages: The heat can build up around the magnet.

Vented: Often referred to as a ported enclosure, this is a sealed enclosure with a port that is tuned to a frequency. There are two types of ports: (1) round and (2) slot (square), and the slot port tends to sound better than a round port. It's a good practice to have the inside and outside edges of the port slot rounded or flared with sand paper.

Advantages: The port improves the low-frequency response of an enclosure. It is extremely efficient and will cause a boost of volume around the tuning frequency of the port.

Disadvantages: They won't play below the tuning frequency without distorting and can become damaged if you were to do so for an extended period of time.

Band-Pass: An enclosure that is similar to a sealed and vented enclosure, but it is the most complex to build. The difference is that the woofer mounts inside a chamber and the chamber fires into a vented box. The chamber and the vented box make up an entire vented box.

Advantages: They have a lot of power and a really clean low-frequency response.

Disadvantages: They are hard to build and all of your dimensions must be calculated precisely.

Free-Air: Also called an infinite baffle, this enclosure utilizes the car's trunk as the enclosure. The subwoofer mounts on a constructed baffle and the baffle extends across the area behind the back seat of a car and is sealed to the car, thus using the trunk as the enclosure.

Advantages: This is the absolute cheapest way to build an enclosure. This also saves trunk space because you won't have a big subwoofer box sitting in the trunk.

Disadvantages: They are complex to build and handle the least power. They also don't handle low-frequency response very well.

The first thing you need to do when designing a subwoofer enclosure is to decide which type of enclosure you want to build. As you can see from the subwoofer information above, each type has its own set of advantages and disadvantages, so it is really up to you as to which one will work best for your skill level, vehicle size, vehicle use, and budget.

Once you've picked your enclosure type, you need to take a measurement of the area you have to work within. You need to measure the depth, height, and width of the area in your trunk where you think your box will fit; this measurement will help you find the cubic feet of air space for your box. For example, let's say you have a trunk space of 28 inches wide, 10 inches high, and 18 inches deep. Multiply these three numbers together and then divide them by 1,728. This is the formula for cubic air space. Thus L (28) x H (10) x W (18) = 5,040. Then 5,040/1,728 = 2.91. So if your subwoofers require 1.5 cubic feet each for air space (the required air space volume for the woofer is found on the box it is sold in), then you know you need 3 total cubic feet of air space for your enclosure.

If you were doing a sealed box, then the 2.91 cubic feet in our scenario would be close enough for your two subwoofers. However, for any other type of box, you must be exact in your measurements. Always subtract the width of the wood you're using for your box from your measurements because the cubic air space should only be the internal measurement of the box, or the air space within the box. You also need to subtract 0.5 cubic foot from your total box volume as an average displacement of the woofer. These measurements are vital to the success of your enclosure.

If your subwoofer requires a smaller amount of cubic air space and your area to build in is larger, then you want to reduce the measurements of the enclosure until they match the requirements for your woofer. You always need to have an enclosure that supports the needs of the subwoofer.

After purchasing your subwoofers, you need to find the efficiency bandwidth product (EBP) of the driver. EBP = fs/qes, where fs is the driver's free-air resonance in hertz. This is the point at which the driver's impedance is at maximum. The qes is the driver's Q at resonance (a measurement of the control coming from the speaker's voice coil and magnet). The fs and qes values are usually provided in the specifications included with your subwoofer. A general rule is that an EBP value of 50 or less equals a driver that would perform better in a sealed enclosure. EBP values of 50 to 90 will give you the option of using a sealed or vented enclosure. Finally, an EBP of 90 or higher is going to work best as a vented enclosure.

If you're building a ported enclosure, you may want to consult a speaker/box building software program (such as Bass Box Pro or WINISD) to help you configure your port. All ported openings have specific measurements and calculations that must be done, so it helps to use the software until you become proficient at doing these calculations yourself. I recommend tuning your port low, somewhere between 31 and 33 hertz.

Subwoofer Enclosure Construction Methods

The most popular choice of material for building a subwoofer enclosure is MDF, usually in a minimum size of 3/4-inch thickness. Your front baffle should be thicker, preferably two pieces of 3/4-inch MDF. You could also opt to use acrylic, ABS plastic, or thick laminate to build your boxes. I have even heard of enthusiasts who have used granite to build their speaker boxes.

In addition to your material, you must be sure to make your box is very strong. When you are finished, you should be able to park your car on top of your subwoofer enclosure without it being destroyed. (Don't actually try this, but you get the idea.) Use plenty of wood glue on every inch of every seam, and press your joint together until the glue squeezes out. Keep a wet rag around to clean up the excess glue. Also be sure to use an internal bracing of MDF wood, as well as generous amounts of wood screws. By predrilling your MDF before placing the screws in the box, in addition to counter sinking them, you can keep the wood from splitting.

Another helpful tip for building an enclosure is to use binding posts through the wall of the enclosure where the speaker wire enters the box. This connector allows you to drill a precision hole that is easy to seal around. You attach your speaker wire to this post on the inside of the box with a ring terminal, and then attach the wire coming from the amp to the post on the outside of the box. These connectors can be purchased from Parts Express (www.partsexpress.com) for $4–$5 per pack.

Finally, be sure to put a gasket seal around where the subwoofer mounts to the enclosure. You can buy this type of weather-stripping at any home improvement store.

Step 1: When constructing your subwoofer enclosures, be sure to use plenty of wood glue so that when you press your edges together it squeezes out on the sides.

Wet Rag

Step 2: Use clamps to hold your subwoofer box enclosure pieces together until the glue dries. Be sure to keep a wet rag handy for wiping the excess glue that squeezes out of the joints.

Step 3: Be sure to use wood screws (not nails) where all of your panels join together. Predrill the holes with a small drill bit first, and then use a countersink to open up the hole just where the screw enters. This will allow the screw heads to sit flush.

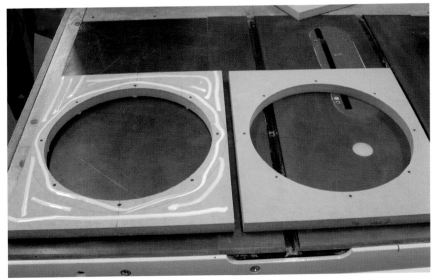

Step 4: Several pieces of wood can be joined together to make the enclosure even stronger in the area around where the subwoofer mounts. I recommend doing this in the area where the speaker mounts because this is the point at which all the energy originates. Be sure to use heavy-duty polyurethane glue for joining the pieces together.

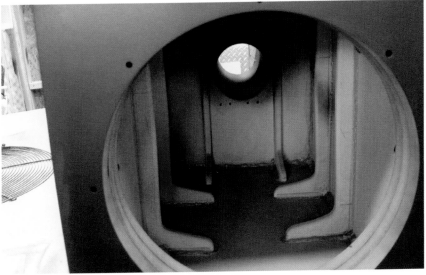

Step 5: You should always add internal bracing inside your enclosure for extra strength and durability. This can be achieved by using small, cut pieces of wood.

Step 6: Binding posts should be installed in the enclosure wall. This allows the signal coming from the amplifier to make its way inside the enclosure. Be sure to connect your speaker wires securely with lock washers and nuts to the binding posts. This photo shows the inside speaker wire connection to a binding post.

Step 7: Here you can see how the binding posts appear from the outside of the subwoofer enclosure.

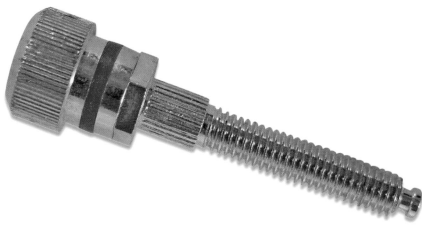

Step 8: In some applications, a traditional binding post will not make its way through the extra thick walls of a sub enclosure. Sometimes you will need an extended binding post for use in these thick boxes. In this picture, you can see an example of a longer binding post made by Dayton, which are available for $5 a pair at Parts Express.

Step 9: Always install a weather strip or gasket on your subwoofer before mounting it into its enclosure.

Installing a Subwoofer Inside a Car

The most important thing you need to do when installing a subwoofer in a car is to make sure it is secure. A secure subwoofer serves three purposes: (1) it won't vibrate unnecessarily and cause rattling and other excess noises; (2) it will not come loose while you are driving and pose safety risks; and (3) it can prevent theft.

I once installed a subwoofer enclosure so securely in a car that car thieves were unable to remove it. I'm not kidding—they stole the entire car, stripped it of every inch of car audio electronics and engine parts, but they had to leave the subwoofer enclosure behind with the car's frame because they just couldn't get it unfastened. Let's take a look at how you can do the same type of installation in your car. We'll start with a discussion about fabricating a subwoofer enclosure, and then I'll show you how to mount one. For this discussion, we'll look at a 2001 Dodge Ram.

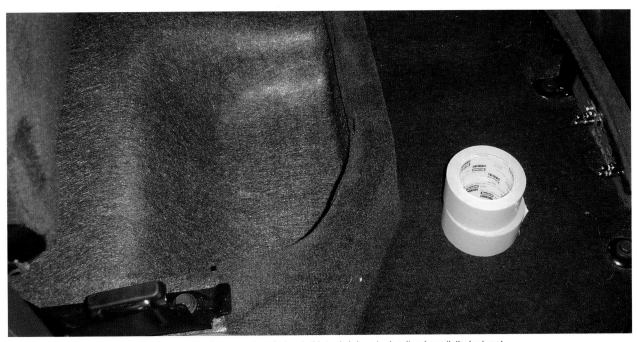

Step 1: First you need to determine where you want to place your subwoofer box. In this truck, I chose to place it underneath the back seat.

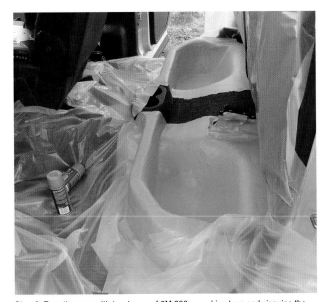

Step 2: Tape the area with two layers of 3M 233+ masking tape and visquine the surrounding areas to protect them from resin and other products.

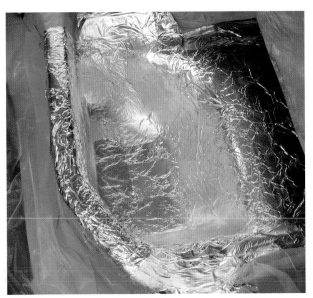

Step 3: Adhere aluminum foil to the masking tape with 3M Super 77 spray glue.

Step 4: Set up a fiberglass workbench to hold all of the tools and ingredients you will use to make your mold. This should include fiberglass resin and mat, paintbrushes, rollers, and a bucket.

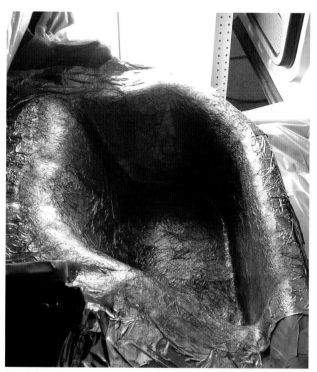

Step 5: Apply fiberglass resin and mat to the interior of your mold using a paintbrush and roller.

Step 6: Remove the fiberglass mold from the car after it has dried for several days, and peel the masking tape off of the back.

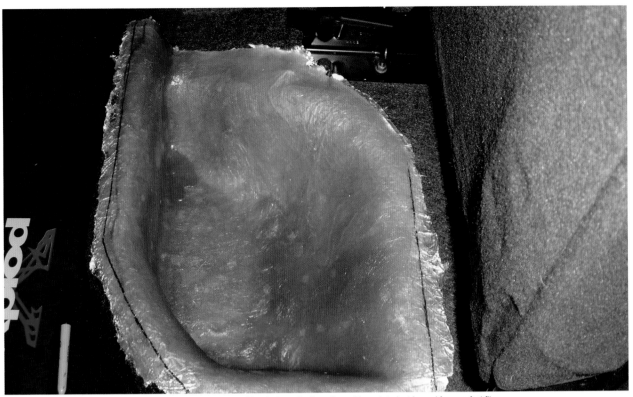

Step 7: Set the fiberglass mold back inside the car and use a black Sharpie to mark where the mold needs to be trimmed for a perfect fit.

Step 8: Trim the fiberglass to the line markings with a jigsaw set to a slow speed.

Step 9: Create an MDF template for the top of your box inside the vehicle and mark the mounting location for the subwoofer. To help you get a perfect fit for the template, start by tracing the template on a sheet of poster board with a black Sharpie marker. Then cut the poster board with scissors and work with the poster board until it fits perfectly. Once it is an accurate template, transfer the poster board to your MDF wood and cut it with a jigsaw. After the shape has been cut into the wood and it fits properly, trace the subwoofer hole and cut the opening.

Step 10: Turn your template into a wood cutout and then cut a mounting hole for the subwoofer; bond it to your fiberglass mold with Duraglas.

Step 11: Once you have bonded the top panel to the mold with Duraglas, sand down the edges until they are smooth.

Step 12: Spray paint the enclosure black to seal the MDF top.

Step 13: Create a cardboard template for a flush top panel that will trim out the subwoofer so you will not be able to see the enclosure.

Step 14: Take the cardboard template and transfer it to 1/2-inch MDF. Drill out the subwoofer mounting holes. Be sure to add spacer braces made out of MDF strips if you need them to level the top of the box.

Step 15: Test-fit the enclosures to make sure that they fit.

Step 16: Drill through the bottom of the enclosure and the floor of the vehicle with a 1/4-inch drill bit. You might want to check under the car first to make sure there is only sheet metal where you're drilling. You don't want to hit any electrical or fuel lines when you drill. Insert a 1/4-inch bolt through your drilled hole, and bolt the enclosure down with washers and nuts.

Step 17: Glue factory OEM carpet to the MDF top and staple it to the edges with an air stapler.

Step 18: Attach the speaker wire leads to the subwoofer before mounting the sub into the enclosure.

Step 19: Hand-tighten the screws for the subwoofer into the predrilled holes that will hold it to the enclosure. At this point, your enclosure should be mounted securely and ready to rock.

Installing Component Speakers

Component speakers are typically mounted in the front doors or kick panel area of a car. However, they can be just as effective when mounted in the rear of your vehicle, so long as they are in addition to a set that has been mounted in the front of the car. I'll show how this can be done with a 2004 Honda Civic.

Step 1: Remove the back deck assembly from the rear of your car.

Step 2: Remove the screws that hold the factory speaker to the back deck.

Step 3: Create a bridge-mount for the aftermarket tweeter out of back strap material or plumber's tape. Both can be found at a local hardware store.

Step 4: Mount your replacement component speaker to the deck with the same screws that held the factory speaker in place. Use one of the screws to also attach the tweeter and its bridge mount. Once done, just reattach your rear deck, and the work is done!

Another popular way to install component speakers is to mount the component speaker in the door or kick panel and the tweeter separately in the A-pillar. Shown here, the tweeter and included basket were directly mounted to the plastic A-pillar.

Chapter 8
Source Unit Installation

We already discussed some of the basic wiring techniques that are involved in source unit installations. However, there are a lot of other factors to consider when you begin installing different types of source units. There are so many types of source units and so many types of vehicles to install them in that you need to be aware of some of the challenges that you might face.

FACTORY SOURCE UNITS

A lot of newer cars are being designed so that it is more and more difficult to remove the factory source unit. Factories are going to great lengths to integrate their source units into as many of the dash components as they can. For example, they link the source unit to the air conditioner, the GPS, or even an On Star system. A lot of source units now also come with iPod connections and satellite radio systems. Because of this, you may want to consider leaving your factory system installed, and use one of the following options for connecting your amplifiers:

1. JL Audio developed a product called the Clean Sweep. This product wires into your OEM source unit and gives you the ability to connect amps without any loss of sound quality. For more information, visit www.jlaudio.com. This product accepts the audio from the source unit and outputs it to your amplifiers with a high-quality transfer.

2. You can take the speaker wires from the factory source unit and wire them into an amplifier that has a high-level input. Make certain if you do this that the factory speaker wires are not amplified, or you will have amplified signal coming into your amplifier.

Single DIN Source Units

Single DIN source units are usually either basic AM/FM CD players or they're DVD players. Remember what we discussed in Chapter 5? When installing a source unit, start by measuring your DIN size and then purchase your wiring kit, harness, and source unit.

Step 1: Single DIN installations generally consist of a mounting cage as the primary bracket to hold the source unit.

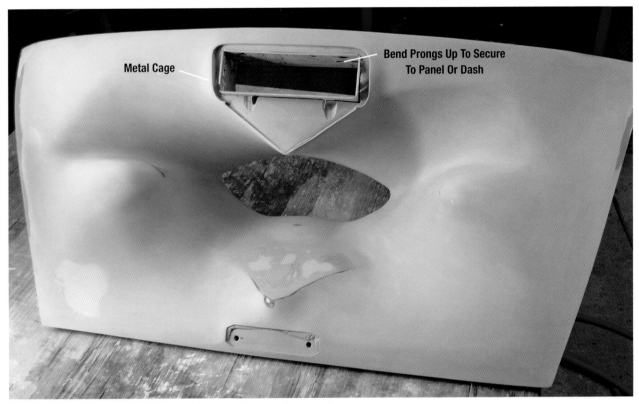

Metal Cage

Bend Prongs Up To Secure
To Panel Or Dash

Step 2: Insert the metal cage into the DIN opening (whether the dash or other fiberglass trim panel).

Step 3: Bend the metal prongs up to hold the cage in place. As an alternative method, shown in this picture, attach metal brackets to the side of the source unit to bolt it into place, instead of using the cage and prongs.

Step 4: Prewire the source unit by running the power, ground, and RCA cables up through the cage prior to sliding the source unit into place. For more information, see Chapter 5.

Step 5: After plugging all of your wires in, slide the source unit into place until it locks. At this point, your installation is complete.

Step 6: Here is another example of a single DIN installation, this time on a Jeep Liberty. Notice the black plastic around the source unit. This belongs to a kit that was installed in the dash first, before the source unit was pressed into place.

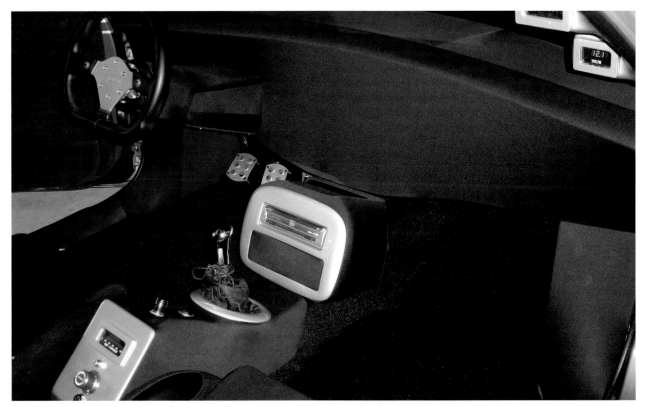

Step 7: This aftermarket source unit's mounting cage and kit were molded into a custom piece of fiberglass. The mounting face formed the front of a custom center console, which positioned the source unit very close to the shifter so that the driver and passenger could both access it easily.

DIN and a Half Source Unit Installation

DIN and half is the most common type of source unit location size. What I show here is how to install a single DIN source unit into a DIN and a half sized location. This installation was done in a 1998 Chevrolet truck.

Step 1: The factory slot is a DIN and a half for the factory source unit. This unit is removed to make way for an aftermarket single DIN source unit.

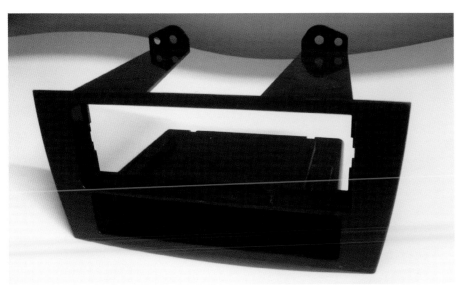

Step 2: When placing a single DIN source unit in a DIN and a half space, you will need to buy a DIN and a half kit.

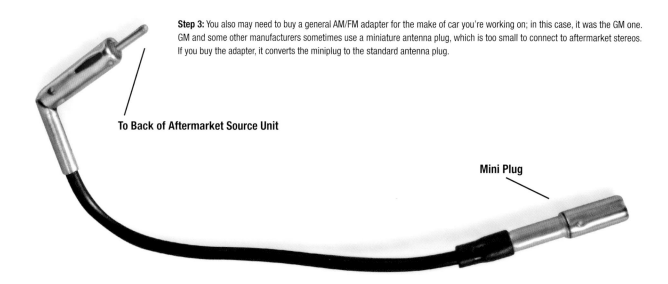

Step 3: You also may need to buy a general AM/FM adapter for the make of car you're working on; in this case, it was the GM one. GM and some other manufacturers sometimes use a miniature antenna plug, which is too small to connect to aftermarket stereos. If you buy the adapter, it converts the miniplug to the standard antenna plug.

To Back of Aftermarket Source Unit

Mini Plug

Step 4: Remove the plastic dash trim in the area around the factory source unit. Squeeze the clips on the sides of the unit to help it pop out from the location.

Step 5: Test-fit the aftermarket wiring harness to the Molex plug that was unplugged from the factory source unit. This picture shows the wiring harness. You should always test-fit these to make sure they fit before you go too far with your installation.

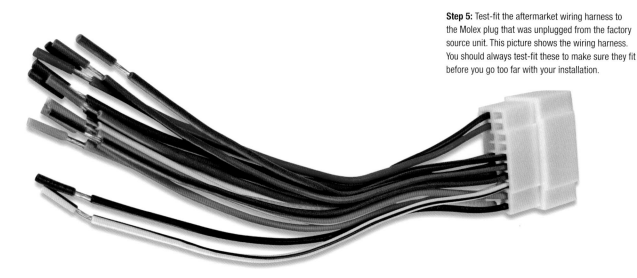

Step 6: Here, you can see an aftermarket Sony source unit that is one of the best sounding source units ever released by Sony.

Step 7: The source unit was unboxed, and all of the included parts were spread out on a workbench and organized prior to beginning the installation.

Step 8: The aftermarket antenna adapter should connect to the back of the source unit.

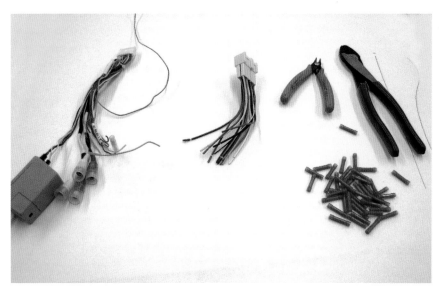

Step 9: Set up a workbench where you can work on the installation. This setup should include the wiring harness included with the source unit, the aftermarket wiring harness, a pair of cutters, a pair of crimpers, and butt-splice connectors.

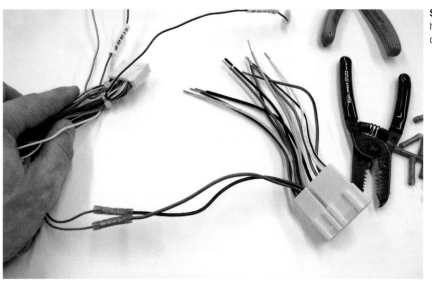

Step 10: Connect the wires between the two harnesses, color for color, and crimp the butt splices on each end.

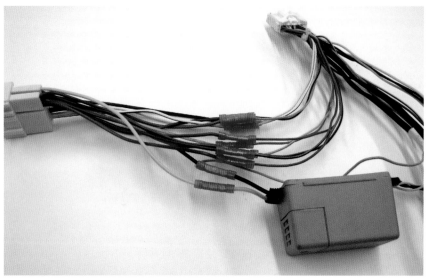

Step 11: Once all the wires are connected with butt splices, the unit is ready to be installed.

Step 12: Slide the aftermarket trim kit into the dash and the metal cage included with the source unit. Bend the mounting tabs to lock the cage into the kit.

Step 13: Slide the RCA cables, factory Molex wire, and the antenna adapter through the opening in the cage prior to connecting anything. Then plug the aftermarket wiring Molex into the factory Molex plug.

Step 14: This picture shows the aftermarket wiring harness connected to the source unit's wiring harness. The Molex plug is ready to be connected to the Molex plug connector in the dash.

Step 15: Plug the remaining wires into the back of the source unit. RCA to RCA, antenna adapter to antenna plug, and Molex plug into the back of the unit.

Step 16: Slide the source unit into the new opening until it locks into place.

Step 17: As you can see, the kit has adapted the single DIN source unit into the DIN and a half opening so that it fits.

Custom In-Dash Source Unit Installation

A custom source unit installation means that you're installing the unit without a kit or a cage. This can be necessary to install a source unit in a custom show car or somewhere where you don't want the source unit to be stolen. Let's take a look at how I did this in a Civic show car.

Step 1: For a custom installation, you can use a 1/4-inch aluminum plate and through-bolt it into the side of your source unit as a mounting option. This will give you an extremely solid mount.

Step 2: I test-fit the console and the source units to make sure that they fit inside the car.

Step 3: In this case, the source unit is being placed into a custom-fabricated center console, so an area needs to be built around the source unit to hide the units and display only their faces. To do this, I cut MDF ribs out of 3/4-inch wood and glued them together to create the outer housing for the source units.

Step 4: I made wood templates to act as trim rings around both of the source units' faces.

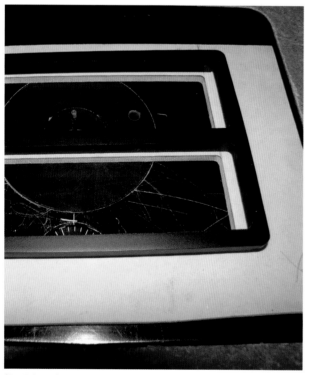

Step 5: I transferred the MDF templates to ABS plastic and routed them to the proper shape.

Step 6: This is the completed ABS trim ring.

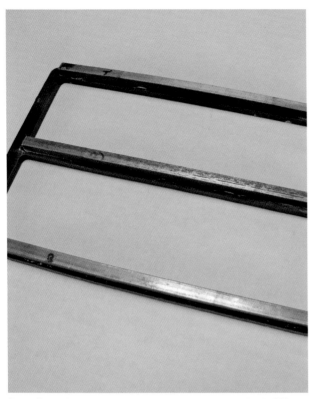

Step 7: Quarter-inch steel square stock is cut to shape and bonded to the ABS plastic to keep it from flexing.

Step 8: Black grille cloth was used to cover the console, and the source units were wired and installed.

Step 9: Once the console was completed, it was installed in the car. As you can see, it houses two single DIN source units.

Chapter 9
Amplifier Installation

Most amplifiers are relatively flat, so there are a lot of creative places you can find to mount them. Some people mount their amps in an inconspicuous place in their car, while other people built flashy amp racks to display their equipment. In this chapter, we're going to take a look at both installation methods so you can decide which type will work best for your system.

SAFELY MOUNTING AN AMPLIFIER DIRECTLY TO A CAR

My all-time favorite place to mount an amplifier—if in a daily driver car—is underneath the seat. This mounting location is very secure because a potential thief cannot see the amplifier easily. It is also extremely challenging for a thief to get to the amp because he/she would have to remove the seat to get to it. Under the seat is also a good location because it keeps the amp cool and allows it to ventilate. In the pictures below and on the next page, I will show you how to safely mount amps under the seats in a few different types of cars.

Building a Custom Amplifier Rack

If you decide to showcase your amplifiers with a fun and flashy amplifier rack, then your imagination is your only limitation. The trickiest part is applying practical engineering and design to your imaginative ideas. I have built some crazy amp racks, and I remember points in time when I spent hours deciding how to convert my crazy artistic drawing into a supportable and attainable design.

A custom amp rack should have a base of steel and aluminum. I have found it useful to shop at local scrap stores to find cheap materials. These types of stores tend to have assorted shapes and sizes of metal pieces that can end up being the perfect fit for your amp rack design.

The sheet metal that supports the seat rails in the car is a structurally sound mounting point for safe amp mounting. This amp is mounted into the floorboard with 8x32 bolts in this Civic.

Two amplifiers are mounted under the seat in this Nissan truck.

Always check to make sure the seat can slide over the amps, unobstructed, without grabbing any wires.

Once you build your frame, you can begin to add shape to your rack with MDF wood and fiberglass. Amp racks can be painted to match the exterior of your car or covered in fabrics to match the interior of your car. However you choose to do it, don't let your funky design ideas become pipe dreams. I have never thought up an amp rack that I wasn't able to build. Remember, a custom amp rack can really become the focal point of the design of your system. You can even give it a theme. So don't become discouraged with the magnitude of building and work that it might take because the end result is always worth the challenge.

The following pictures show a few of the amp racks I have built over the years. Each one had its own set of challenges, which I will try to spell out for you so that you can understand some of the factors you'll need to consider when you begin to design your own amp racks.

This is an amp rack that holds four amplifiers in the back of a Jeep Liberty. The challenge with this rack was that I had to come up with a design that I could build, sand, and paint all in one night.

This is an amp rack in the back of a Civic show car. The rack holds nine Sony amplifiers, and it was a challenge to fit the rack in the back of a two-door car. I had to make the rack in three pieces so it could be moved in and out of the car without getting scratched or stuck.

Believe it or not, these two amp racks were the front driver's and passenger's seats in a Civic. They each held three amps, for a total of six. Seats had to be functional for the sound judges to sit comfortably on them. Another challenge with these racks was finding a way to make them mount safely to the car without damaging the fiberglass. I was actually once forced to drive this car on the Las Vegas Freeway while sitting on these amp rack seats.

This amp rack holds nine Sony amps and takes up the entire back seat of a Civic coupe. The back seat is not a functional seating area because of this rack, but it does make a cool display area for some video monitors. This rack was built in three separate pieces, which had to be carefully seamed together with fiberglass molding.

This amp rack holds six Sony amps and also serves as the rear seat of a Civic hatchback. This particular amp rack weighs several hundred pounds and is only one piece of fiberglass. Because of this, taking it in and out of the car was a three-person task to ensure that it didn't get dropped or brushed against another part of the car. Unless you have a lot of experience with custom amp racks, I do not recommend building one all-in-one piece. *Joe Greeves*

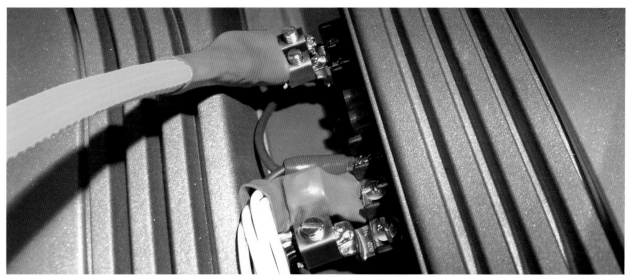

Flat ring terminals, made of copper, provide a clean connection for the amplifier because they allow you to secure the wires to the terminal with a setscrew, and you only have to connect the ring part of the terminal to the amplifier.

Amplifier Terminations

You need to properly terminate your amplifier wires so that they are clean and safe with the proper connectors and connections. This way, a strand of copper doesn't jump and touch the terminal next to it and short out the amp or your speakers. Here are some examples of really clean amplifier wiring.

TIP!

When running your cabling for your amplifiers from the front to the back of your car, run the power cable on one side of the car and the RCA and speaker wires on the other side of your car. This will keep extraneous noise out of your system.

All signal and power wires should enter the area around the amplifier separately from each other. You don't want a power cable running directly next to an RCA if at all possible because it can generate noise in your system.

Heat-shrink tubing should be applied in the area around where the wires enter the amplifier terminal. Black rubber or plastic grommets protect the wire as it passes through metal.

Notice how the wiring for these six amplifiers is laid out in a neat, uniform way.

This amplifier termination photo shows the power cables coming in from one side and the RCA cables entering from the other side of the car.

Be sure to connect all of your accessories and car electronics to the same ground point to prevent ground loops from entering into your system. This picture shows multiple amps and batteries grounded to an aluminum bus bar that is connected to a rear strut mount.

Shown here is an 8-gauge power wire connected to the terminal labeled 12 volts. This wire should come off your fuse distribution block. Be sure to disconnect the battery from the car before connecting this wire.

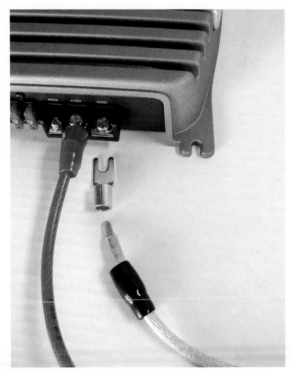

Next, crimp another brass fork terminal to the silver 8-gauge ground wire before connecting it to the chassis ground terminal.

AMPLIFIER INSTALLATION

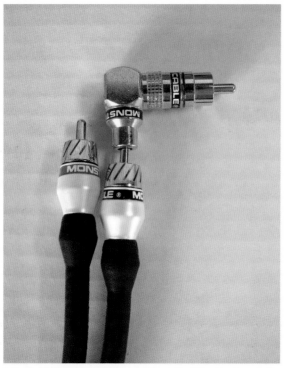

A Monster Cable right angle adapter can be used for RCA cables if you don't have a lot of room to plug your RCA cables into your amplifier.

You should always securely mount your amplifiers to your car. I prefer not to use any screws but rather to use bolts. Shown here is an amplifier mounted with 8/32-inch x 1-inch bolts.

Another option is to build your own RCA cables. This is probably one of the most overlooked options for most installers. Building your own RCA cables can be cheap because it is just a piece of wire and a connector. The connectors shown here are made by Neutrik, and you can get them at Parts Express for about $20 for a pair. In this photo, the shiny brass section in the center is where you connect the positive wire, and the duller brass section on the outer edge is where you attach your ground wire. You repeat this connection type at the opposite end of the cable, and that makes the other end of the cable.

Shown here are the adjustments for sound that you can find on most amplifiers. The one on the left is level, and this is the gain adjustment and will adjust the overall volume of your system. The one in the center is low boost, and I recommend doing nothing with this except to turn it all the way down. The one on the right is the filter, which is the crossover point adjustment. If you use the amps on subwoofers, you turn this on; if you use the amp for midrange, you leave this off for this specific Sony amplifier.

Amplifier Setup and Adjustments

When terminating an amp, the very first thing you want to do is ground the amplifier. The amplifier needs to have the heaviest gauge wire that can fit on the terminal run down to a seatbelt bolt (see Chapter 5). Then you'll want to connect all of your amplifiers to the same ground point. If you're using an external crossover and EQ, be sure to turn the amplifier's internal crossover off. You don't want the two types of crossovers to interfere with each other. A ground loop can occur when you have products in the car grounded to one point (such as amplifiers in your trunk grounded in the trunk), while you have other products grounded at another point (such as a head unit in the dash grounded to the dash). This creates a ground loop, which will induce noise into your system.

Another important part of amp setup is turn off the bass boost—I have never seen the need for this function. If you're using your amp for a mono connection, be sure to switch the internal switch to mono.

Setting the gains on the amplifier is often an overlooked step in car audio systems. A simple way to do this is after you have your system connected and playing music, set all of the gains on low. Then turn the source unit volume to about 95 percent of full volume and slowly turn up the gain (sometimes called sensitivity) while listening for the sound of the speaker to start distorting.

This is often a two-person task. When the speaker starts to distort, stop and turn the gain back down until the distortion goes away. This will be the point to where your amp is sending the maximum output that your speaker can handle.

Installing a Capacitor

A capacitor should be wired in-line between the battery and the subwoofer amplifier. It should be wired as close to the amp as possible, within an inch if you can manage it. Try to use a piece of copper tubing or copper wire to get the capacitor terminal directly next to the amp terminal. Remember, a capacitor does not need to be fused, but the power wire coming off of the battery has to be fused.

You only need a capacitor on your subwoofer amplifiers. Only on very exotic systems would you think about placing a capacitor on your other speaker's amps. You can connect multiple capacitors together—positive to positive and negative to negative—for extra capacitance for your amps. Be sure to follow the wiring instructions provided earlier in this book for charging a capacitor. Capacitors are not dangerous if you charge them with a resistor prior to installation. Let's take a look at some examples.

This capacitor is wired directly next to the amp's positive connector terminal, using a copper bus bar that has been ground down to the desired shape. This makes the close connection possible.

Aluminum or copper bar stock can be used to connect multiple capacitors together. Be sure to connect the polarity correctly—all positives together and all negatives together.

Be sure to secure your capacitor firmly to the car using the supplied mounting bracket, which will usually come with your capacitor.

Chapter 10
Video, Crossover, and Equalizer Installation

As I have mentioned before, video, crossovers, and EQs add the flash and finesse to a car audio system. The video can add both flashy lights and entertainment value, while the crossover and EQ can tune your audio system and make it perform at its optimal levels. I invite you to take a tour of some installations that I have done with each of these electronic components.

CROSSOVER AND EQ PLACEMENT

As we discussed earlier in the book, a crossover and an EQ should be placed in-line between the source unit and the amplifiers. This is because they adjust the signal from the source and send it where it needs to go through the amps. If you do not have an EQ or crossover, it's okay because many amps have built-in crossovers that you can use, and many source units have built-in EQs. Keep in mind, however, that the built-in crossovers and EQs do not perform as well as the external ones.

When mounting a crossover and EQ, try to remember that you will have to be able to adjust them frequently. Also, the initial setup and adjustment of these components may take several hours. You do not want to get in and out of the car constantly to adjust a crossover or EQ in the trunk, so mount it as close to the listening position as possible.

Some car stereo companies that produce EQs and crossovers have a remote display that plugs in with a phone cable (RJ-11), and this display can be mounted in the front of your vehicle, making remote tuning much easier. In these cases you can mount your EQ and crossover pieces wherever you want to in the car.

Try to mount your crossovers and EQs close to your amps so that your cables can be really short. This helps to minimize the cost of your cables, as these can really add up in price. For example, if you mounted a crossover in the trunk and an EQ under the front seat, you would have to run 12-foot cables, which are four times the price of 1-foot cables. One 12-foot cable could cost you more than $100.

Crossover and EQ Adjustment

Subwoofers mounted in the trunk should be crossed over relatively low. A good point of reference is about 60 hertz and below. I usually cross over my front stage speakers to be around 80 hertz if I am using a two-way system.

A two-way system describes front speakers with trunk-mounted subwoofers. A three-way system would add midbass drivers up front. In a three-way system I would cross over a trunk-mounted sub to play 60 hertz and below, the front midbass to play 60–150 hertz, and the front midrange to play 150 hertz and above.

One thing you may find with crossovers is that they want you to set a slope for the crossover points (see Chapter 1). A general slope is around 12 dB per octave for your midranges. Steeper slopes seem to work better for your midbass and subwoofers.

Some crossovers and EQs connect to the source unit with a fiber optic digital cable. This provides the absolute best sound quality.

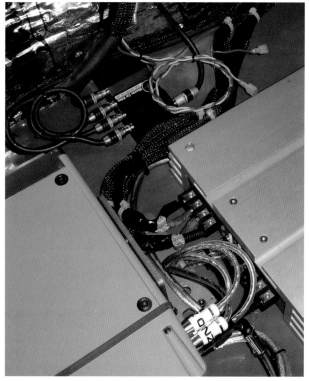

Shown here is a crossover/EQ connected to an amp with really short RCA cables. Not only does this save you money in cabling, but it provides much better sound because the waves have a shorter distance to travel. There is less chance of noise getting into these wires.

Shown here is a power and ground connection for an audio processor with an EQ and crossover built into it.

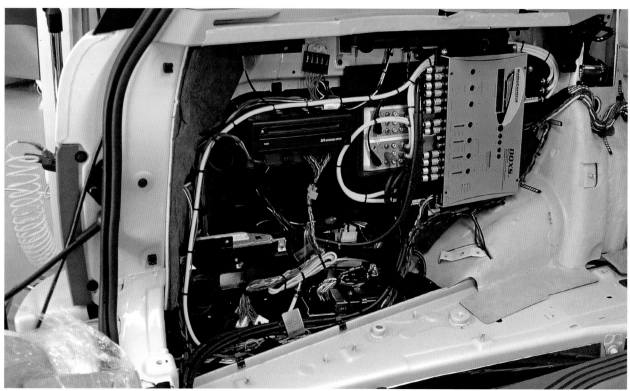

This picture shows detailed wiring of an audio crossover and an EQ in the back of an SUV. This trunk-mounted installation was convenient because there was a remote display mounted in the front of the vehicle for making adjustments.

Adjusting EQs can be done by ear, or with a real time analyzer (RTA) machine. An RTA tells you in a graphical format the acoustical frequency response that your car's system is playing. An RTA is a great way to learn how to tune your car until you can hear it by ear, but unfortunately, one can cost around $10,000. If you don't want to spend this kind of money, most car stereo shops will tune your car with an RTA for a small fee.

As a recommendation for adjusting your EQ settings, I suggest cutting the frequency levels rather than boosting them.

VIDEO SAFETY

When installing video monitors in a car, it is essential that you don't put them in front of the driver. The video monitors in the rear of the vehicle can be viewed by passengers while driving down the road, but you cannot view a monitor from the driver's seat. Please follow all local and state laws regarding video monitor installation.

If you are installing an overhead video monitor up on the headliner, it is important that it is securely fastened to the car. I recommend a support bracket with bolts. You do not want the video system to come crashing down to injure someone during a collision.

VIDEO COMPONENT INTEGRATION

Most people want to integrate video monitors into their daily drivers either in the headrests of the seats or from the interior roof as a flip-down. Both options provide convenient video viewing from the back seat without distracting the driver up front. If you have kids, then you already know what a lifesaver a mobile DVD player can be on longer car rides. Let's take a look at a simple way to install video monitors into factory headrests.

INSTALLING A VIDEO MONITOR

In today's busy world, you may find that you want a ready-made video installation kit. You can now buy aftermarket headrests with video monitors already installed in them to match your car and its interior. These kits can save a lot of installation time, but they don't always provide the most custom look, so you have to weigh your priorities before purchasing this sort of kit.

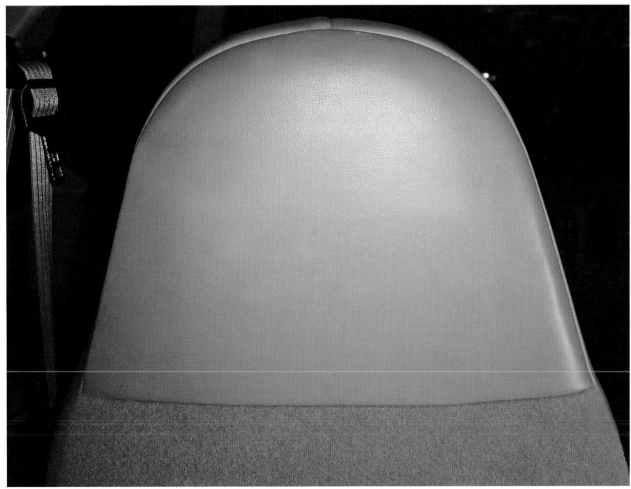

Step 1: Evaluate the back of your seat to see if the video monitor will fit within the workspace, such as between the seams.

Step 2: Here is the video monitor that will be installed into the headrest. There will be two monitors, one in each of the headrests.

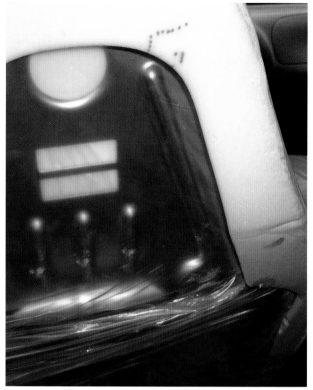

Step 3: Remove the seat cover and see what existing material can be found inside the seat that can be used to mount the monitor.

Step 4: Your monitor should come with an ABS plastic housing to mount the monitor in place. Use ¼-inch MDF to create a template of the housing.

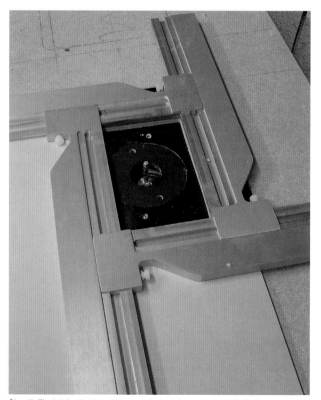

Step 5: Flush-trim the template on an inverted router using a Fakuda router tool as a guide.

Step 6: Test-fit the ABS housing in the template.

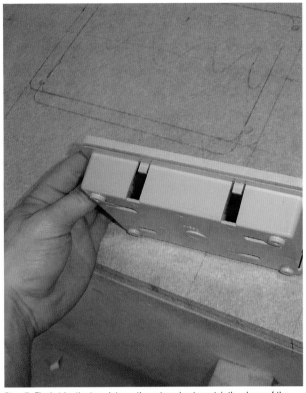

Step 7: Flush-trim the template on the outer edge to match the shape of the ABS housing.

Step 8: Transfer the template to ½-inch MDF to build up the monitor templates to the correct thickness, then flush-trim it on a router.

Step 9: Glue the templates together and then test-fit the monitor housing. Also be sure to mark the opening for wire access on the template.

Step 10: Notch wire access holes into the template using a router and a cutting bit.

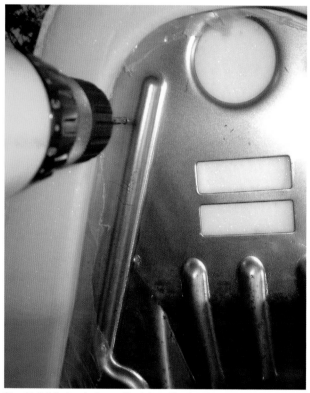

Step 11: Drill the housing's mounting bolt positions into the back of the seat's supporting metal in preparation for the mounting bolts.

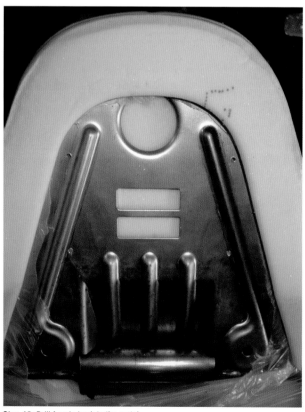

Step 12: Drill four holes into the metal.

Step 13: Bolt the monitor housing to the seat with four 8x32 bolts.

144

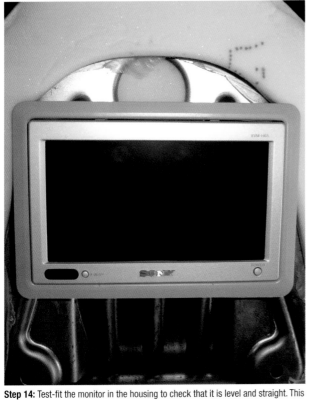

Step 14: Test-fit the monitor in the housing to check that it is level and straight. This is very important, as you do not want a crooked video monitor!

Step 15: Use a metal bender to make four custom brackets out of 1/8-inch flat stock. These will reinforce the mounting position.

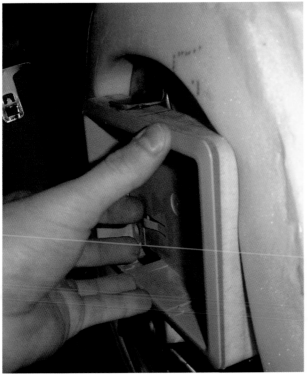

Step 16: Bond the brackets to the template and bolt them through the top of the seat. The bracket is glued to the MDF, and then that structural frame is through-bolted to the seat.

Step 17: Prewire the video connecting cable into the housing and zip tie it into place to keep it secure during installation.

Step 18: Feed the video cable down through the bottom of the seat and zip tie it to the metal underneath the seat. Be sure to keep it out of the way of the seat slider so that the cable doesn't get damaged.

Step 19: Reinstall the seat cover over the top of the new baffle and cut the fabric to allow the monitor housing to sit flush inside the wood frame. A Sharpie marker can be used to trace the impression in the seat material of the baffle. Then you can run a razor blade over the Sharpie outline to cut the fabric.

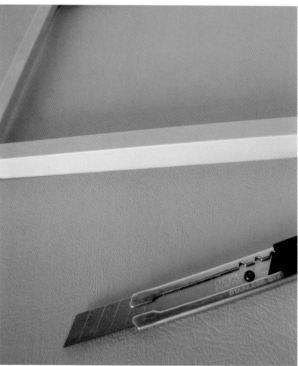

Step 20: An extra monitor housing can be used as a guide to cut straight lines in the seat fabric.

Step 21: Shown here, an Olfa stainless-steel knife is used to cut the seat opening for the monitor.

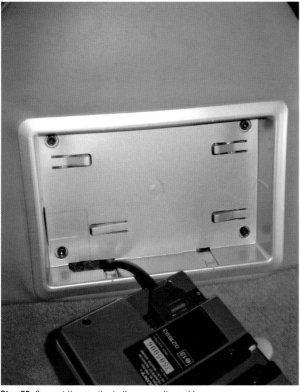

Step 22: Connect the monitor to the connection cable.

Step 23: Snap the video monitor into the housing to complete the installation. For this entire project, allow yourself two days.

Step 24: As you can see, the finished product gives a symmetrical view of both monitors.

Step 25: An in-dash, single DIN DVD player is perfect for running video screens in the headrests.

Video Setup

Once you have installed your mobile video units, it's important to calibrate the monitors. This should be done on two settings: (1) daytime viewing that is calibrated outside while in the bright sun and (2) nighttime viewing that is calibrated when it is dark outside. These settings should be saved on different presets to provide one setting for daytime and another for night. Do not attempt to do this while driving down the road.

Here, you can see six LCD monitors' power supplies double-stacked so that they only take up the space of three. A 1/4-inch piece of ABS plastic is used as a mounting point.

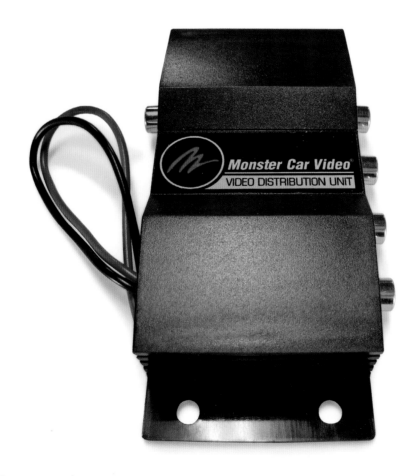

If you are installing multiple monitors and you're daisy-chaining the video signal, then this can cause signal degradation. Monster Cable makes a signal booster for monitors, and it's designed for 12 volts. In this picture, you can see one of these boosters displayed.

This is a Sony XAV-A1 touch-screen DVD player used in conjunction with a Sony CDX-GT805DX. The touch screen functions only as a DVD player, and the source unit provides all of the radio and CD functions. It is possible to have the touch screen operate individually, and then it would maintain its DVD functions but could also be used for radio and CD.

Here, you can see a factory double DIN source unit with CD and cassette in an Infinity.

The factory source unit has been removed from the same Infinity, and the AC controls were relocated lower. A Sony source unit and 7-inch LCD screen were installed where the double DIN unit was located previously.

Here are four LCD monitors installed in the back of a Honda Civic to highlight the installation and the amplifier rack.

The LCD monitors in the same Civic were installed flush, in individual tribal fiberglass stands.

This is a custom heads-up display on a headliner that houses three 7-inch monitors.

HOW TO MAKE A CUSTOM CROSSOVER FOR A TWEETER

Sometimes tweeters intercept unwanted bass. In order to prevent this, you might want to create a custom crossover for the tweeter. This will protect it from distorting or blowing because of the bass.

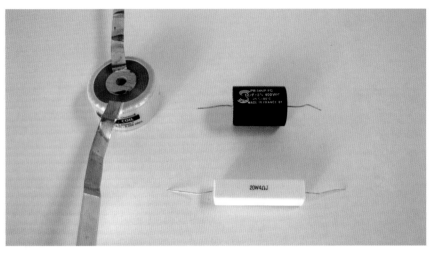

Step 1: These are the traditional parts of a passive crossover network. On the left side is an air-core conductor, or coil, which consists of a piece of wire wrapped around a wooden, hollow center piece to create resistance. The black object is a polypropylene capacitor made by Solen. At the bottom of the picture is a sandstone resistor.

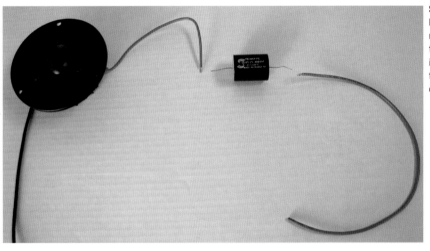

Step 2: These are connections on a tweeter. The black wire goes to negative on the amplifier, and the red wire goes to positive on the amplifier. But before the red wire connects to the tweeter, you must install a capacitor in-line between the amplifier and the tweeter. This capacitor creates a 6 dB crossover on a tweeter.

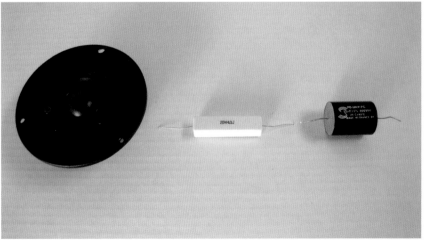

Step 3: A lot of times the tweeter will still be too loud when wired to an amplifier with just a capacitor on it. Then you need to add a sandstone resistor on the same line on the positive lead to tone the tweeter down a little bit. A 4 ohm resistor is shown here. This will drop the tweeter about a 1 1/2 dB.

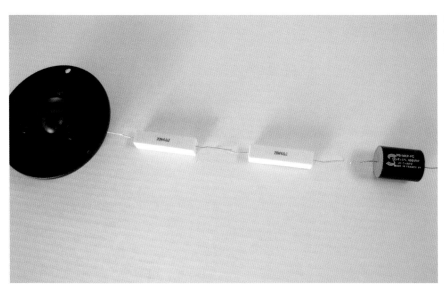

Step 4: Here is another resistor wired in-line in case the tweeter is still too loud. This is what is called passively tuning a tweeter.

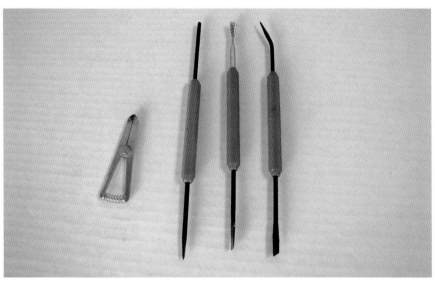

Step 5: Soldering tools can be purchased from Sears Craftsman for about $3. On the left is a heat sink, which is designed to absorb the heat and keep it off the electronics when soldering. The other tools are a pick tool and a metal brush (for cleaning the connections before soldering).

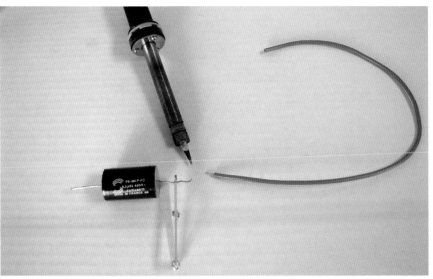

Step 6: The capacitor is soldered to the tweeter wire. You always put the heat sink next to the closest component.

Chapter 11
Custom Installation

Once the last electronic device is mounted, the last wire is wired, and the final component is tuned, it is time to start adding the artistic touches that will make your system a unique expression of your personal style. You can do this in a number of ways, such as adding accent lighting (as we previously explored), building custom door panels, creating trim panels, using fiberglass to build mounts for the electronic components, doing body work to your car, or even using custom paint both inside and outside of your car. Your imagination (and your budget) is the limit, and I invite you to dream big.

Creating artistic masterpieces with fiberglass is one of my most cherished crafts, and I am glad to share some of the tricks of the trade with you in this chapter.

CUSTOM DOOR PANELS
Custom door panels are one of the hardest types of custom installations, and many installers struggle to make their designs function well. This is partly because they can get heavy with all of the fiberglass and equipment that you might want to put into them, making the doors sag on their hinges. It also because once you begin to build out a door panel, the door is no longer the same width or depth as it started, so it may begin to interfere with the doorsill and not close properly. You can also run the risk of it hitting the dash, causing damage to your doors and your dash and keeping the doors from closing all the way. These are just a couple of the challenges that you must plan for when starting a custom door panel project.

Step 1: Remove the factory door panel and tape up the inner door skin with double layers of masking tape. Do not apply this to tinted windows! The blue tape at the top of the door panel is painter's tape, and it is safe to use on window tint.

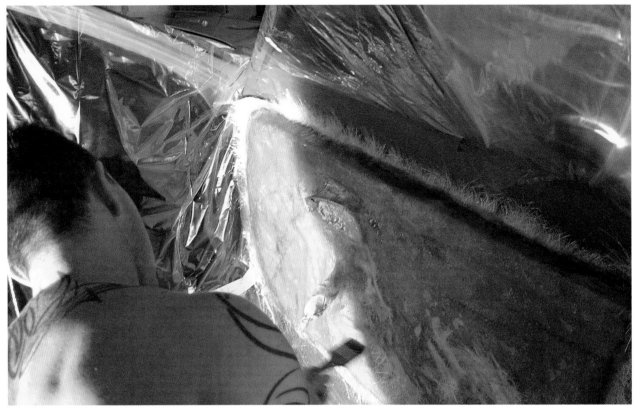

Step 2: Apply six layers of 1.5-ounce fiberglass mat to create a base mold.

Step 3: After several days' drying time, remove the base mold from the car, mark the outer edge with a Sharpie marker, and trim it with a jigsaw.

Step 4: Tape white poster board together and cut it to create the custom shape of the main section of the door panel.

Step 5: Trace the poster board onto ¾-inch MDF to create the main section of the door panel shape.

Step 6: Trim and block the fiberglass mold with a sanding block until it is the desired shape that accommodates your custom design.

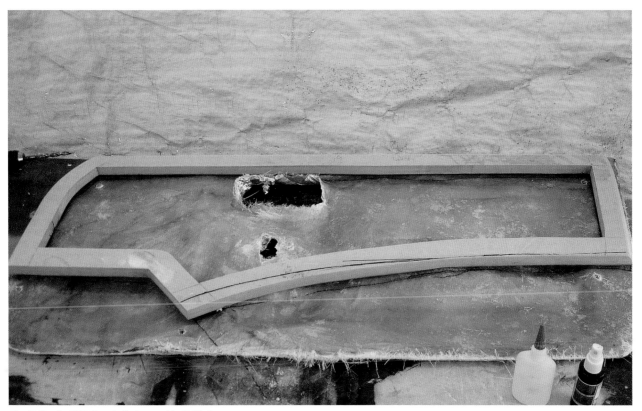

Step 7: Glue the MDF shape to the base mold with CA glue.

Step 8: Create an MDF template and rout it to give the lower door panel shape.

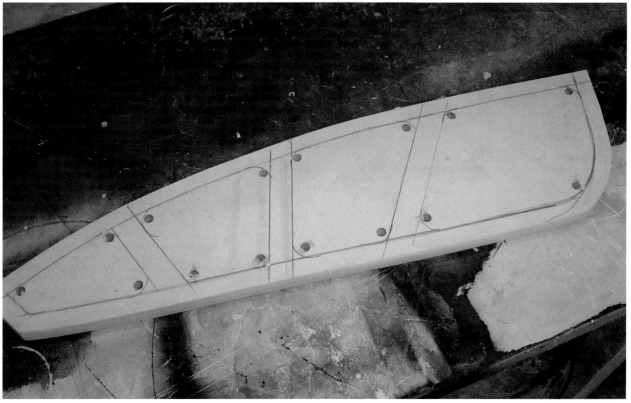

Step 9: Trace the design onto the wood and drill pilot holes for trimming with the jigsaw.

Step 10: Test-fit all of the MDF shapes on the base mold of the door panel.

Step 11: Thread the wiring for the door controls through the base mold shape. This wiring should be able to pass through unobstructed.

Step 12: Create a cardboard template for the shape required where the door panel meets the dash. This must be very precise!

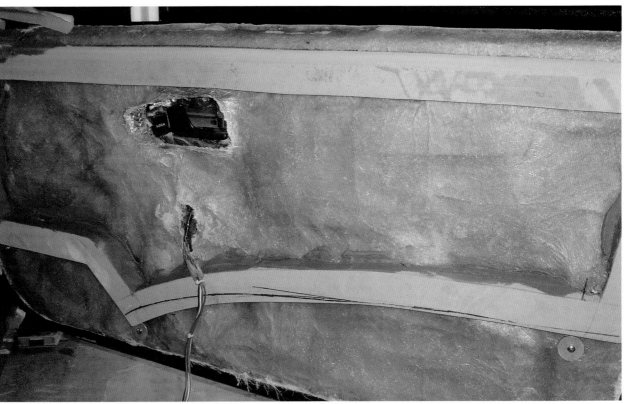

Step 13: After a trial fit, bond the MDF template to the mold with Duraglas.

Step 14: A 1/2-inch piece of MDF wood is bonded at the top of the panel with CA glue and then Rage Gold body filler is spread evenly on top of this wood lip and around the top edge of the door panel. It is then sanded smooth to create the door panel shape.

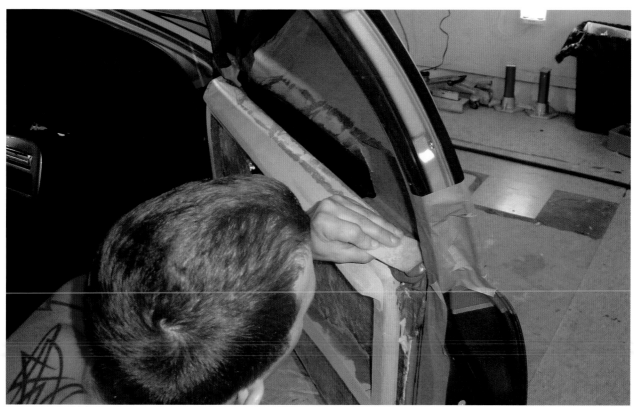

Step 15: Apply Duraglas on the edges instead of Rage Gold because it gives extra strength with less chance of chipping or breaking.

Step 16: Wrap the door with a fabric of your choice. In this case, I used black grille cloth. Make sure that the fabric is stretched closely over the mold.

Step 17: Here is a finished picture of another custom door panel that housed three 6 1/2-inch speakers and three tweeters.

Step 18: Here is a finished shot of a custom door panel with an optical illusion built in: Notice the futuristic concept car molded into the base of the door. The speakers act as the tires for the car. *Joe Greeves*

Step 19: This is the same door panel but with three different speakers inserted into the door panel.

CREATING CUSTOM TRIM PANELS

Most of the time, a custom trim panel can be made using a factory part. This is a very easy way of achieving a custom look without having to spend the time or money building a one-of-a-kind part. In this short tutorial, we'll go over building custom A-pillars to trim out tweeters, using your factory plastic parts.

Step 1: Mark the tweeter placement on the A-pillar with a Sharpie marker. Make sure when you do this you consider the size of the tweeter—if it is placed too close to the windshield and it can hit even slightly, you will have an annoying rattle.

Step 2: Choose the appropriate ABS plastic mounting ring that will work with your marked position on the A-pillar.

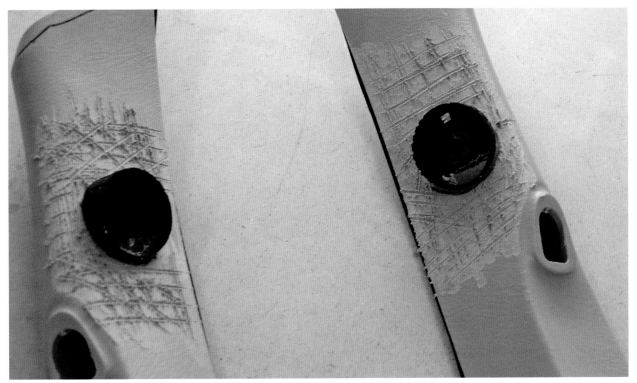

Step 3: Drill a hole for the tweeter wires to feed through. Glue the tweeter buckets to the A-pillar, and then scuff the A-pillar plastic with a die grinder. Then cut notches in the plastic with a razor blade.

Step 4: Tape the tweeter buckets with masking tape and apply Duraglas to the A-pillar, building it up evenly around the tweeter bucket.

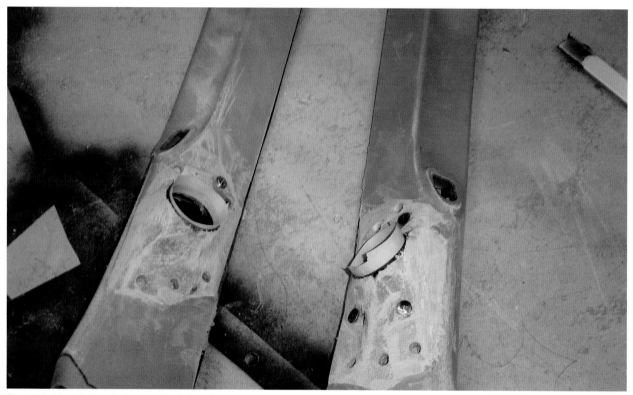

Step 5: Grind down the Duraglas in preparation for body filler.

Step 6: Apply body filler to both A-pillars.

Step 7: Sand down the body filler so that the A-pillars are ready for texture paint.

Step 8: Spray paint the A-pillars to match the interior color with SEM (www.sem.ws) paint, then reinstall them in the vehicle, and connect the wires. Allow yourself at least two days for this entire project.

FIBERGLASS AND BODYWORK AND AUTOMOTIVE PAINT AND CLEAR COAT

When creating custom fiberglass pieces for your car, there is a certain level of sanding and bodywork that must be done on a panel to get the finish to look like glass. I am often asked how I get my panels to look so much like glass. Doing this type of fiberglass and paint is not easy, and it takes a lot of practice.

I would recommend to anyone who is a beginner to take a vocational course in paint and bodywork from a local community college to learn some basics techniques. These courses are often not expensive and are really worth the money and the time. If you have a friend who already knows some of these techniques, then buy him or her a six-pack and settle in for a few weekends in their garage.

Let's take a look at how to get a glassy finish with fiberglass and paint.

TIP!

To ventilate a garage for paint, place box fans in the garage doorway all the way across the opening and blowing outward. Close the garage door until it sits on top of the fans. Open any windows to allow air to blow in, but make sure there are screens to block bugs from flying into your paint. Wear a respirator to protect your lungs. You should try to have as much air intake as possible when working in a small space.

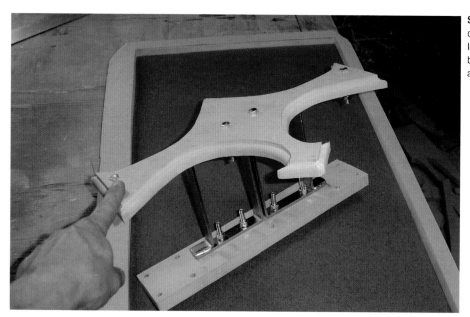

Step 1: Use 3/4-inch MDF and cut it into your design preference for what you want to build. In this case, the wood was cut to design the base of a custom seat. Steel brackets were added to the MDF to hold it together.

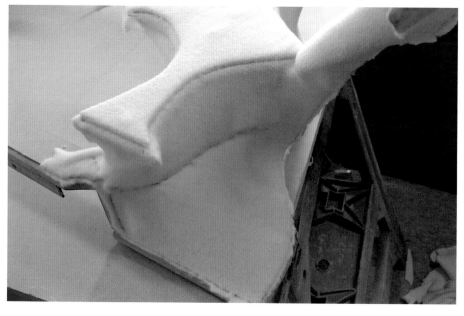

Step 2: Stretch fleece over a wood frame and staple it to create your shape.

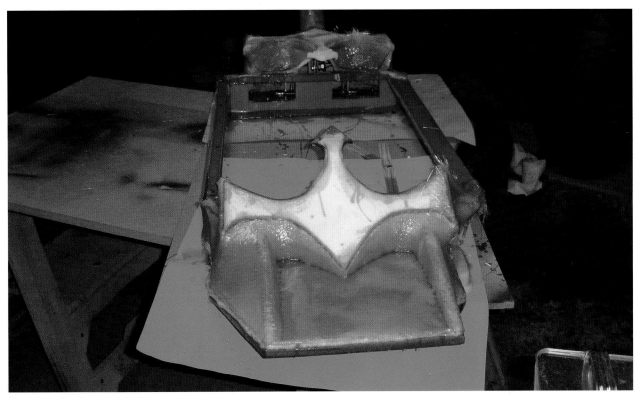

Step 3: Apply fiberglass resin to the fleece to create a mold.

Step 4: Allow the mold to dry, and then trim the fiberglass to shape.

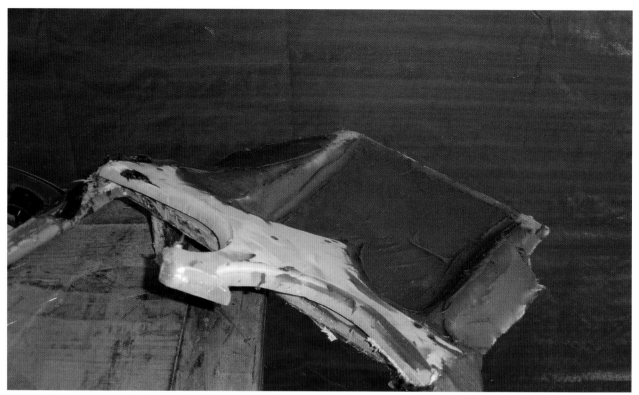

Step 5: Spread Duraglas body filler evenly on the panel.

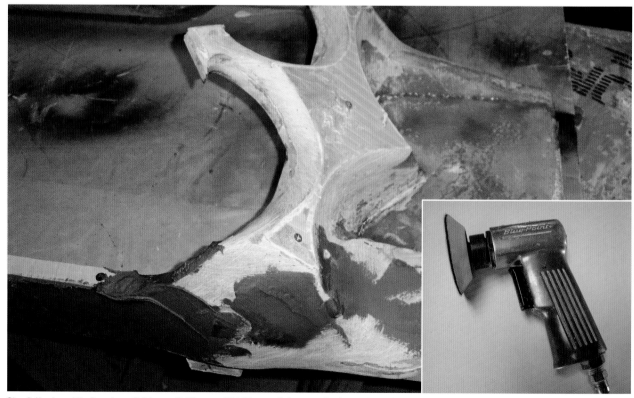

Step 6: Hand-sand the Duraglas until it is smooth. When you think it is smooth, keep sanding for another half-hour. (Inset) Shown here is my favorite tool for customer installation, a 3-inch DA sander made by Bluepoint. This tool is excellent for sanding body filler and doing bodywork.

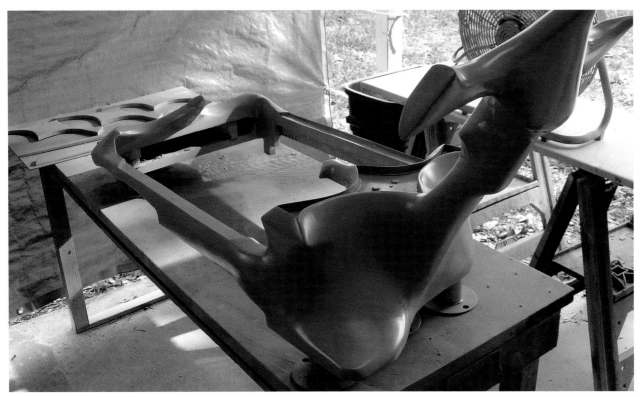

Step 7: Once you have achieved smoothness, spray several coats of high-build primer and allow it to dry. When working with paint and other chemicals, wear protective gloves and clothing and try not to expose your skin to the chemicals.

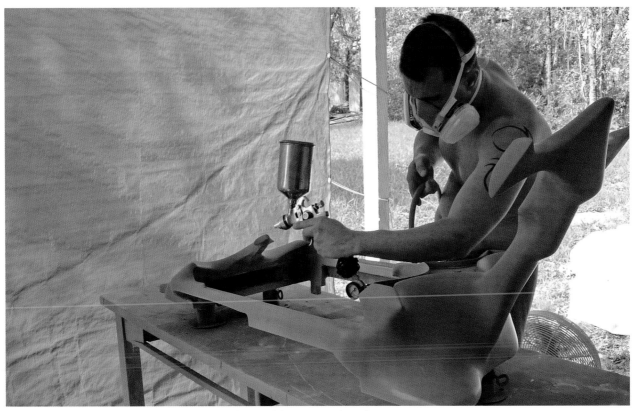

Step 8: Wet-sand the primer until it is smooth with 220-grit sandpaper, and then you can apply your base coat.

Step 9: Do not think that you have to have a high-end shop or expensive paint booth to have a quality finish on your painted parts. The pieces shown here were painted inside a home garage. Notice the infrared heat lamp—this helps the paint dry quickly. Make sure you use proper ventilation, and wear a respirator.

Step 10: Shown here are the completed seats.

CREATING SYSTEM HIGHLIGHT PIECES FROM EXOTIC MATERIALS

Any time you want to highlight a certain portion of your system's installation, there are a variety of choices: Aluminum, acrylic, and ABS plastic can be used to create some really interesting pieces. They can be machined on a router table just like wood to conform to any shape.

You must use caution when routing exotic materials. Rough-cut your piece into a desired shape (with a jigsaw) that is within a few millimeters of your marked line before putting it on the router. That way the router has to trim very little off of the piece to make it perfect.

Step 1: Cut an MDF template of a trim ring for an ignition and wiper switch. The finished product is going to be aluminum, so trace the template onto the aluminum, using a metal scribe.

Step 2: Flush-cut the inner aluminum openings on a router.

Step 3: Cut the inside opening on the router to a 45-degree angle with a chamfer bit.

Step 4: Flush-trim the outside edge of the aluminum on the router. Do not worry about scratching the aluminum, as it will be polished later.

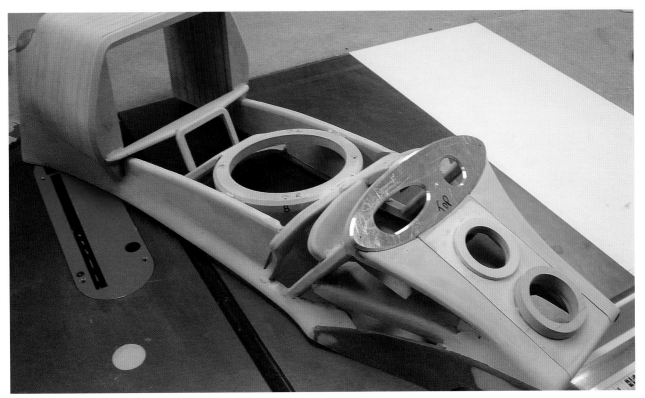

Step 5: Test-fit the piece to the area where it will be mounted, in this case a custom center console.

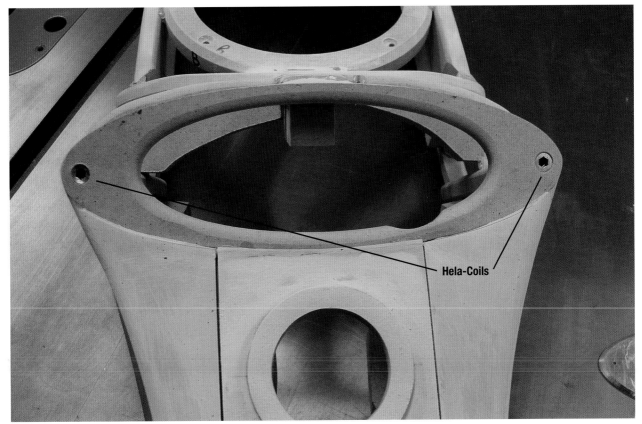

Hela-Coils

Step 6: Inset Heli-Coils (threaded metal inserts) in the wood as a mounting point for the aluminum.

Step 7: Polish and install the piece, and the installation is complete.

Here is the finished 1/4-inch aluminum trim bezel with the edges routed.

CUSTOM UPHOLSTERY WORK

Some people choose to wrap their fiberglass molds in fabric rather than painting them. Other people use custom upholstery to stitch mats, seat covers, and door panels. When planning to cover something in upholstery, you must plan for the material thickness you will add to your piece. So when designing, leave enough of a gap on the edges to allow for the thickness of the material plus glue. If you don't do this, your wrapped pieces will not fit into their assigned places.

Another cool technique is to use a rabbeting bit to cut a groove in the area where the excess material will wrap on the edge. This slot will provide a place for the material to go and keep it from bunching up on the edge, causing your panel not to fit.

The adhesive glue that I recommend for use with fabrics is Wellwood Contact Adhesive. It's best to spray this on the back of the material and on the surface to which the material will be applied. I usually try to apply the material after the glue has tacked up for about 10 minutes or so. After the glue is tacky, I start applying the fabric a small section at a time while maintaining awareness of the amount of fabric I am using. If you work in small sections, it is easy to use more material than you planned, and then when you get to the end, you are short on fabric. You must be vigilant when stretching your fabric while working in sections. If possible, have another person help you hold the fabric while working from section to section.

If you don't have a person to help you, you can use a thin layer of spandex to divide the fabric from the surface you are laying it on. Spandex does not stick to the underlying glue, so you can lay your excess fabric over the spandex while you adhere the rest of your fabric to the surface you're working with.

Finally, always be sure that you drill mounting holes for your subwoofers or amps before you lay out your upholstery.

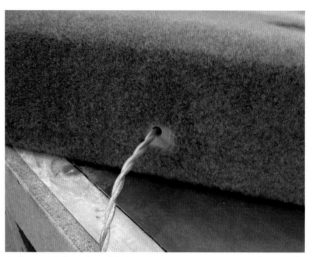

Apply carpet to the surface of the box. It's a good idea to notch the carpet where the wires penetrate the box, so the silicone that seals the wire will adhere to the wood.

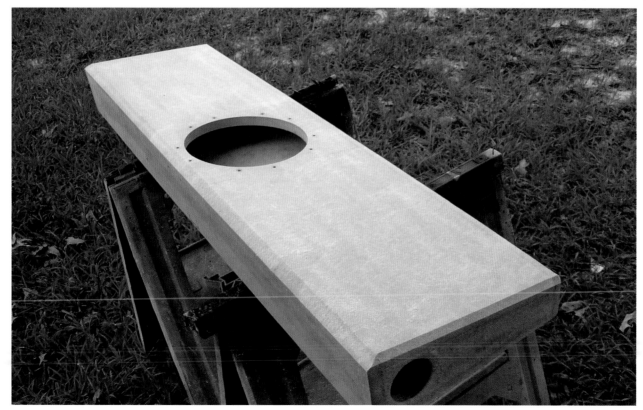

Spray adhesive should be sprayed on an enclosure in preparation for covering it with carpet.

This is a finished, carpeted subwoofer enclosure. Notice that the face was carpeted first so there are no visible seams; the seams are brought around the box and put out of sight.

Here is a completed, carpeted subwoofer box.

SHOWCASE

Here are some miscellaneous pictures that showcase some of the custom projects I have worked on over the years.

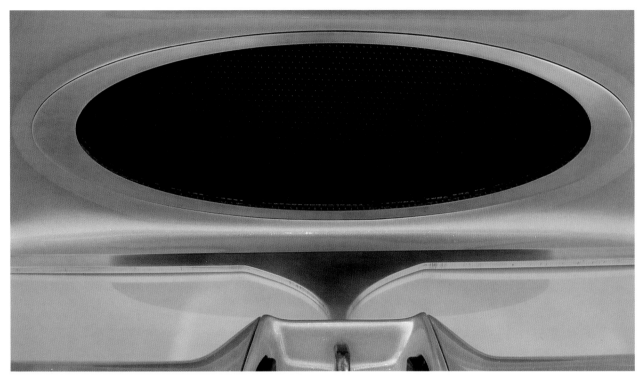

This is a custom-machined aluminum opening around a subwoofer port.

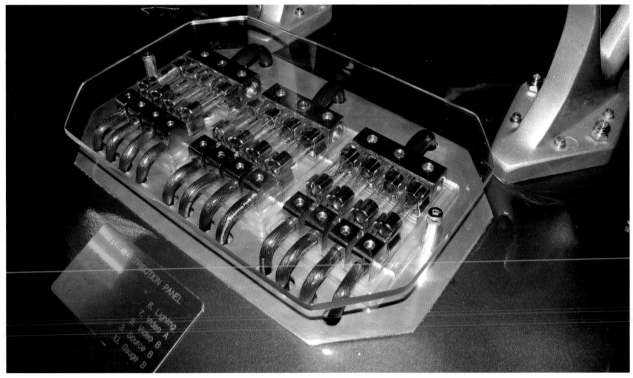

This is a custom power distribution panel that has three fused powered distribution blocks.

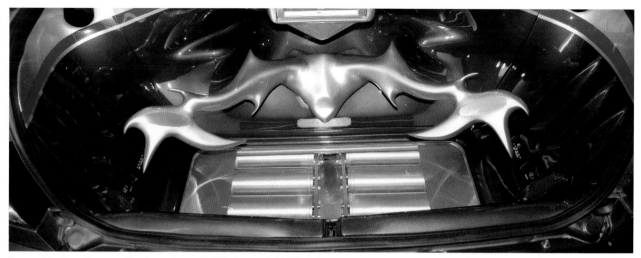

If you want to get creative, you could make a custom, molded fiberglass trunk with LCD displays to monitor battery voltage and a capacitor back with eight capacitors, as shown here.

In this picture, paint on the exterior of the car flows through the door jambs and seamlessly onto the fiberglass panels of the interior of the car. This is a lot of work for a painter and must be done precisely. Few people take on this project because of the amount of labor involved, but the effect is definitely worth it.

Here, you can see a custom tweeter rail on the top of a door panel that holds two Sony 1-inch dome tweeters.

A custom paint job can be a great way to highlight your entire stereo installation. It takes time and money to get a good paint job, but it can really be the component that takes you car to the next level. *Joe Greeves*

Chapter 12
Car Shows and Competitions

If you're a mobile electronics enthusiast, then you've probably been to a car show or two. For those of you who are working hard to fabricate, build, wire, and tune an awesome custom system, you might want to consider showing off your hard work for others to see.

There are two ways you can go about showing your car. You can take it to a car show just to show it, often with the sponsorship of a vendor, magazine, or other company; or you can take it to a competition car show and compete with it. You can compete for any number of titles: bass, sound quality, paint, craftsmanship, motor/engine work, or best in show. The sound quality competitions judge your car's sound system based on balance, tonality, stage, loudness, and installation integrity. Often sound quality competitions include judging SPL (sound pressure level), flatness of sound (with an RTA machine), and creativity of design.

I've shown and competed with my cars professionally for almost two decades. Looming car shows often were the deadlines that helped me to finish projects and usually meant a crazy rush of busy days and sleepless nights. There are a lot of things you need to know before you can just travel off to a car show, and I hope this chapter will help you get prepared.

A car show is a great way to draw a crowd around your vehicle. You can get a large number of people sometimes, so I recommend making "Do Not Touch the Car" signs if you want to protect any custom paint or delicate pieces in your system.

Car shows are both a great way to meet people as well as way to get media coverage for your car. Magazines and television shows are always present at car shows.

COMPETITION ORGANIZATIONS AND RULES

Competing with your car is a lot of hard work. Just getting your car to a point where it is complete enough to take it to a competition is an accomplishment all on its own. There are many different organizations that judge cars for competitions, and they all have their own rule books and guidelines. All car organizations have membership and competition fees, so be prepared to spend a little money. You also will need to factor the cost of travel.

Two of the more popular sound quality and SPL competition organizations are USACi (United States Auto Sound Competition International) and IASCA (International Auto Sound Challenge Association). You can obtain their rule books by visiting their Web sites (www.iasca.com and www. soundoff.org) and becoming a member. Their rules often revolve around the way their logos must be displayed on your car and how your wiring must be finished. They require that you show start-to-finish pictures of the building of your car,

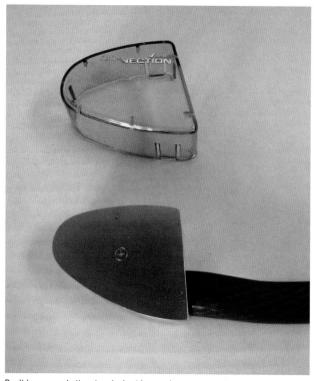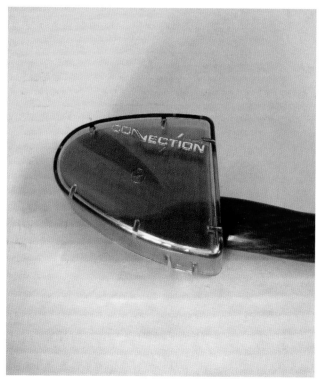

Don't leave your battery terminals at home when you compete your car stereo! These battery terminal covers, made by Audison, are a good choice for covers. They are inexpensive but look really nice.

and you should be prepared to give a full presentation to the judges about your installation.

Once you've obtained a one-year membership with an organization, you get a rule book and a competition CD (by which you'll be judged). The rule book is something you should read cover to cover long before you even begin to build your car system. The rule book explains standards for wiring, speaker placement, installation guidelines, and expectations. However, this doesn't mean that you can't just show up at a car show with a competition going on and enter your car for judging. Even if your system has not been built for that organization, by having it judged on the spot you can learn a lot. This sort of critique can show your strengths and weaknesses and help you determine if competition is a sport for you.

Car audio competitions started growing in popularity in the early 1980s, and the competitions focused mostly on loudness of the system and flashiness of the installation. During that time, it was very common to see gigantic subwoofers (larger than 20 inches) fill up the back of cars. In the early 1990s, organizations began to structure themselves with rules and regulations for the competitions' formats and began to examine aspects of the cars, such as sound quality and installation integrity. As these new organizations grew in size, they began to offer large cash prizes for the winners, as much as $20,000. At the same time, large crowds began to flock to the shows.

Interest in the sound quality competition shows has gone up and down over the years. In the 1990s, the competitions drew hundreds of competitors, but in the earlier years of the millennium the competitor numbers diminished to a few dozen. In recent years, the number of competitors, as well as the number of spectators, has skyrocketed again, and the industry is presently soaring.

So what does this mean for you? Depending on the organization you join, cash prizes have begun to be offered again. You might spend a lot of money building your car and traveling to the show, but the glory of a title and a little extra cash are certainly worth the effort.

Most importantly, car shows are a time to meet and network with people that share your interests and passion for quality mobile electronic systems. I am very fond of what I call my "car show family," which is the people I only see five to ten times a year, at every car show. They are always there with the newest additions to their cars and families, ready to spend a weekend with other friends and enthusiasts. It is more fun than you will ever have elsewhere.

PREPARING FOR A SHOW

Once you've chosen a car show you want to attend, you need to make sure you have your car ready for the show as far ahead of time as you can manage. My advice is to have the car stereo installation completed 30 days before the show. I recommend this, but I have never actually seen

An easy-up tent is a good way to protect your car from the elements of rain and sunshine at a car show, especially if you have custom paint.

it happen. This is because there is something wrong with us car audio guys. We can have a checklist of items that we plan to complete for a show. These items can be scheduled perfectly, and as soon as those items are complete and ready thirty days before the show, we decide to create another list of things to do because we have thirty days to spare, and it will make the car look really cool for the show. Next thing you know, you're on a suicide project to get the car done in time, and you spend the entire week before the show not sleeping so you can finish on time. It all really just boils down to the burning desire to have the coolest car at the show. I will warn you now, by adding this kind of crazy, last-minute workload you are far more likely to lose the event.

Having the coolest car is not the winning answer; the best way to place in an event is to be prepared and finished.

When you're preparing for a show, create a checklist of achievable items on a set schedule. No matter what you do, stick to the list. Finish the projects, but do not add to them. If you finish early, then spend that time cleaning the car, putting together a photo album, and practicing your presentation. I also recommend getting plenty of sleep because there are a lot of fun things to do at car shows after the judging is done.

Once you begin to get close to the 30-day countdown to the show, begin to prioritize your remaining tasks. Try to finish the most critical projects first, and place the less important items as the last tasks to get done. This way, if you don't finish them all, you can have at least the most important parts of your system finished. If you miss your 30-day deadline because your projects run long, then it is critical that you create a prioritized task list because chances are good you're not going to get it all done in time. Make sure you always have little projects you are willing to sacrifice for the sake of time.

Finally, once your installation projects are complete, take some time to go through the show's rule book and make sure that your installation follows all of the rules. For example, if you know that having an open 12-volt circuit, where a judge can touch something conductive, is a five-point deduction, then you wouldn't want to show up to the show with exposed battery terminals. So plan to bring extra battery terminal covers with you to the show. Just as we said in Chapter 3, planning is critical to building a system. So, too, is planning critical for you to perform well at a car show.

As I said, you will be required at most shows to present pictures of your build. This means that you must take detailed and organized pictures of everything you do when you build your cars. They will need to be arranged in an album or slide show that can be examined by multiple people.

WHAT TO PACK

So what do you take with you to a car show? The list is incredibly long. And I will tell you right now, no matter how well you pack, you will inevitably forget several things and end up making countless stops at Wal-Mart along your way to the show.

My best advice to you is to make a list of all the things you may need. Continuously add to this list every time you go to a show and realize what you should have had. I have spent years developing detailed show supply checklists, and I still forget things.

My checklist is the same for all of my shows, and there are two boxes to check next to each item. The first box is to be checked once the item is packed and ready to go, and the second box is for when the item is actually loaded up and ready to go. I highly recommend organizing your checklists,

Another item I use frequently at car shows are ropes to tie off the area around your car. This is helpful if you do not want to have people touching your car, but you must leave it personally unattended for a few minutes to take a break or get some food.

packing, traveling arrangements, and other car show details on a buddy system. If you don't have a spouse or significant other to help you with this, find a trusted friend who is willing to do a little work for a fun weekend vacation. Using a buddy system will help you keep track of details and minimize the number of forgotten items.

Finally, I recommend you map out the town where you are going ahead of time. Find a local car wash where you can wash your car when you get there. Locate the facilities where the show is being held, and find a hotel and some local restaurants.

Things to Pack:

- Photo album
- Entry fees
- Collapsible chairs
- Bottled water
- Easy-up canopy
- Generator and fan
- Cooler
- Extension cords
- Battery charger for the car (so you can run the stereo without the car being on)
- Spare gas (because most convention centers require you keep less than a gallon in the car)
- Camera and/or video camera
- CDs and DVDs for your system
- Emergency money
- The Club
- Floor jack and lug wrench
- Battery-powered impact wrench
- Spare keys and spare tires
- Car cover
- Laptop if your system is controlled by one
- Ropes
- Tie-down straps if you're trailering the car
- Sunscreen and bug spray
- A few cans of compressed air
- Cleaning supplies:

    o Meguiar's NXT Generation Glass Cleaner
    o Bounty paper towels (have less dust and won't streak)
    o Meguiar's Hot Rims Chrome or Aluminum Wheel Cleaner
    o Meguiar's NXT Generation Speed Detailer
    o Meguiar's Hot Rims Mag & Aluminum Polish
    o Meguiar's Microfiber Cloths

- Tools:

    o Socket set
    o Leatherman Wave
    o Crimpers
    o Wire strippers

o Electrical tape
o Primary wire
o Spare fuses
o Volt meter
o SPL meter
o Phase detector
o Test light
o Hand truck (often your parking will be very far from your car's show space)

HOW TO GET THERE

Depending on how advanced your car is, you may not actually drive your show car. I have cars that are daily drivers and cars that are show cars. My show cars only get transported to car shows. However, if you are transporting a car, you have to have the means to do so.

If you can get an enclosed trailer with a big truck rig to pull it, then it can be an ideal way to get to a show. These rigs are often and easily stolen, so make sure you load up with plenty of security. By driving your own rig, you can supervise the car at all times. I recommend getting an enclosed trailer, preferably made by Pace. These can be custom built for your car for under $7,000. Make certain your trailer has interior lights and an exterior light that shines on the loading ramp.

To pull this trailer, you will need to have a three-quarter-ton truck with a V-8 engine. These trucks are usually labeled as 250 or 2500 in the model of the truck. The truck will need to be equipped with a Class 4 or higher weight-rated hitch. It's better to have these installed on your truck by a professional with liability insurance and a good warranty. Your trailer will also need electric brakes. Older trucks will also need an electric brake controller installed in the dash. I have found that Tekonsha is a really good one to use.

Newer trucks (2006 and newer) that are in this weight class can be purchased with a tow package, which usually includes the electric brake controller as well as a transmission and oil cooler for long-distance travel. Towing mirrors are also an added benefit and can be bought as manual extends or power versions.

Some people opt to pay for a transport company to take their cars to shows. I have gone this route a few times as well, and it is convenient if you have to go all the way across the country. But if you take this option, you usually have to have your car finished and ready to go two weeks earlier. It can also cost a whole lot of money. I recommend Innercity Lines above all other transport companies. It is the most experienced and safest, and has excellent customer service. The company also covers your car with a million-dollar insurance policy.

When you travel to a car show, make sure you know where you are going. Take a GPS, maps, and driving directions from MapQuest. You can't have too much help with the navigation—trust me. I have gotten lost and

An enclosed car trailer is a great way to transport your car to and from a car show if you do not want to drive it. Some cars are built beautifully and are ruined by daily travel, so transport by trailer is really the only option.

run late for more shows than I can count. Often, I am so tired from trying to get the car done in time that I am too tired to drive to the show. If you don't have people to help you get there, then pad your schedule so that you have time to sleep. It is always better to get a good night's sleep before leaving then it is to sleep in shifts while out on the road.

PRACTICE MAKES PERFECT

Car show competitions are an adrenaline rush like no other. You work hard for months on end to get to a show where it all comes down to the judges. No matter how good your car is or how well you present it, you are not going to always win. This is by no means an indication of your talent or the performance of your car. There can only be one winner, and if it's not you, chalk what you can up for a learning experience and go have a good time at a car show.

I have gone on car show tours before where my car placed first at a show and then went directly to the next

show where it placed last. At the end of the day, the judges make the final call, and you never know which way it will go. Just learn, practice, and perfect your craft, and keep on trying until you achieve everything you want for your car.

FINAL THOUGHTS

There are a lot of different reasons why you might choose to install a mobile electronic system. You might do it for fun, or you might do it for a job. You may be looking to get magazine coverage or you may be trying to create a cool system to jam with on your daily commute to work. It really doesn't matter why you build your system so long as you enjoy doing it. It may be a headache and it may be hard work, but the reward is in the sound, and it's worth it every time.

I have had a lot of great teachers over the years, and I always want to help where I can. If you want to discuss your car's system, feel free to drop me a line at **info@jasonsyner. com**. I'd love to hear from you.